Managing the Engineering and Construction of Small Projects

COST ENGINEERING

A Series of Reference Books and Textbooks

Editor

KENNETH K. HUMPHREYS
American Association of Cost Engineers
Morgantown, West Virginia

Additional Volumes in Preparation

Managing the Engineering and Construction of Small Projects

Practical Techniques for Planning, Estimating, Project Control, and Computer Applications

Richard E. Westney, PE
Spectrum Consultants International, Inc.
Houston, Texas

MARCEL DEKKER , INC.   New York and Basel

Library of Congress Cataloging in Publication Data

Westney, Richard E., [date]
    Managing the engineering and construction of small
projects.

    (Cost engineering ; 9)
    Includes index.
    1. Engineering—Management. 2. Construction industry
—Management.   3. Industrial project management.
I. Title. II. Series: Cost engineering ; no. 9.
TA190.W48 1985   658.4'04   84-28608
ISBN 0-8247-7417-5

MARCEL DEKKER, INC.
270 Madison Avenue, New York, New York 10016

Current printing (last digit):
10 9 8 7 6 5 4 3 2 1

PRINTED IN THE UNITED STATES OF AMERICA

3-6-86

For Marilynn

Preface

What is the big problem with "small" projects? Why has the world of small-project management become increasingly complex, frustrating, and difficult to manage? Why don't the techniques developed for very large projects help those involved in projects which are small? Are there better techniques for managing these small projects than the methods most organizations use, and, if so, what are those better techniques? Are they practical? How can they be implemented? What about those new microcomputers: can they help?

This book is an answer to these and other related questions. It is based on the author's consulting work and training-course presentations in small-project management given over the past 10 years.

Everyone who plans, schedules, budgets, and controls his own work and the work of others is involved in project management. Effective project-management methods are available to make that part of one's job much more effective. Years ago, while marketing consulting services in project management, it became apparent to the author that there was a whole world of project management which had gone virtually ignored by everyone except those actually working in it. This was an area in which there were many unresolved problems, many unique and difficult constraints, and diverse difficulties involving organizations, operations, people, technology, and economics. This was a management environment far more difficult to handle than that of large projects. This was the world of small-project management.

"Small" projects can involve maintenance, upgrading, revamps, turn-arounds and outages, research, engineering, plant improvements, light construction, or environmental work. They can be capital projects, or expensed as operational costs. A project is considered small

when it does not involve a major investment by the company or when
the company itself is small. For these and other reasons described in
Chapter 1, small projects, and the project engineers who are responsible for them, often suffer from a lack of appreciation and support, as
well as an absence of practical management methods.

The project engineer handling one or more small projects is the primary person for whom this book is intended. Chances are, he or she
has little or no formal training in project management, and few practical methods available in-house. In fact, project management is probably just a part of the project engineer's job and, since it is generally
unsatisfying, possibly not the favorite part. However, since cost overruns and schedule delays seem to speak loudly and eloquently about
the project engineer's performance, the project management part of the
job remains important. In response to this situation, the author's consulting firm undertook to develop and market special methods for
handling the problems of the small project. These methods are based
on the latest proven management concepts for large projects but simplified and adapted as required to fit the small project environment.
Short-cut methods are provided wherever possible.

The concepts and methods for small project management which are
presented in this book are practical and proven. They have been
presented in professional training courses to over 1000 project engineers and managers from such diverse industries as oil and gas, petro-
chemicals, nuclear, utilities, government agencies, consumer products,
mining and minerals, pharmaceuticals, aircraft, tobacco, electronics,
engineering and construction, and health care. Companies from all
parts of the United States, Canada, Latin America, Europe, Scandin-
avia, and the Middle East and Far East have participated in these
courses, and also contributed many ideas and experiences. The methods
have also been successfully implemented in many different companies and
work environments.

This book presents a number of ideas and specific techniques which,
taken together, form a complete and comprehensive project-management
methodology. The contents stress concepts, rather than detail, because the small-project environment varies so much from company to
company that the specific way in which the concepts are implemented
will vary a great deal. Although some of the concepts may seem new,
strange, or even impractical to some readers, it must be stressed that
nothing is presented which is not of proven value in the small project
environment. Since project management rests on the same fundamental
concepts, whether the project involved is large or small, the book also
provides a comprehensive review of the basics which will be useful to
anyone involved in projects. Every relevant aspect of project management is covered in this book in sufficient depth that most readers will
be well equipped to handle any small project and most larger ones as
well.

The revolution in the application of personal computers to business problems has had tremendous implications for those who work with small projects, and a separate chapter is devoted to this subject. Although it is possible and practical to apply most of the methods described in this book manually, the ready availability of microcomputers in most situations makes it possible for any project engineer to have simple and effective management tools at his fingertips. To illustrate some straightforward computer applications, many of the illustrations in this book were prepared using the spreadsheet and graphics of the Lotus 1-2-3, and standard printers.

Intended readers of this book include, in addition to project engineers: managers of the project functions, line supervisors in the field or in engineering or research, as well as anyone who must perform some project management as part of his job. The concepts and techniques presented are intended to simplify the job of project management; to make more time available for the type of tasks for which an engineer or manager is trained and paid. Once grasped, they will inevitably make the project management part of one's job considerably more enjoyable and effective.

Richard E. Westney

Contents

Managing the Engineering and Construction of Small Projects

Part I
Defining the Problem and Elements of the Solution

1
The Big Problem of Small Projects

INTRODUCTION

For those of us who work with them, small projects mean big problems.
Unlike our colleagues working on big projects, we have to cope with
special problems like handling many projects at once, working in a production environment, and being on our own in the work we do. There
are, fortunately, some proven project management concepts and methods which can be adapted to fit the small project environment. Before
defining a solution to the small project problem, we must first define
the problem in the most specific terms, and that is the purpose of this
chapter. Although small projects exist in widely different circumstances, they all share certain common elements in both the problem
and solution.

What Is a "Small Project"?

The word "project" can be used to describe an endeavor in which a
number of tasks are performed in order to accomplish a particular objective. In the usual context of project management, projects are usually undertaken by an organization to achieve business objectives such
as:

Maintenance of production capacity
Increased production capacity
Compliance with environmental requirements
Performance of research for new product development

Provision of engineering, mechanical, or construction services
Profit generation from the project work itself

Projects generally consist of one or more of the following broad tasks:

Engineering (technical specification of the work to be done)
Procurement (provision of the required materials)
Construction (assembly, installation and startup of the facilities)

We can see that the determination of what a "small" project is, is really determined by the environment in which the project takes place. However, we can define some guidelines as to what we mean by a small project. Small projects, in the context of this book, have one or more of the following characteristics:

Cost levels from $5,000 to $50,000,000
Cost levels less than 5% of annual budget for projects
Numerous other similar projects take place concurrently
Labor and equipment resources shared with other projects
The company doing the project is, itself, small

Some examples of small projects are:

Plant maintenance
Maintenance on offshore platforms
Research
Computer system development
Plant addition
Engineering
New product development
Plant modifications or improvements
Light construction
Projects to assure compliance with environmental requirements
Utility system outages

SMALL PROJECTS HAVE BIG PROBLEMS!

Size Belies Importance

Most people involved in project management (PM) would consider the types of projects listed above to be quite straightforward: to be not nearly so difficult as the multi-billion dollar "super-projects" to which so much attention has been paid.

The fact is, small projects can be just as important to the company involved as the larger projects and sometimes even more important.

For example, a "turnaround" project, in which a critical manufacturing or process unit is shut down and overhauled in the absolute minimum time, can have a major impact on the plant's profitability if it takes too long and causes valuable production to be lost. Often the timing of the introduction of a new product is of critical importance, and depends on a number of engineering and construction projects to be completed on time.

So the value of successfully completing the small project can be far greater than the cost of the project itself, and the importance of the small project to the plant should not be underestimated just because the cost is small. If the company doing the project is itself small, the project may represent a major investment to that company.

Small projects are also important because of their increasing cost and complexity. As industries develop, and inflation continues, projects tend to become more complex and costly, making many types of projects suitable candidates for a more sophisticated approach to project management than had previously been taken.

The total cost of small projects is often not small at all. In a large plant, such as an oil refinery or steel mill, the individual small project may represent an insignificant sum, but the aggregate cost of all the small projects done each year may be significant indeed. For example, the cost of maintenance each year often exceeds the expenditures for large capital projects and, unlike a large project which lasts a few years, maintenance work goes on continuously. So, if the total program of small projects is considered, it generally does not represent a small project at all!

Perhaps the most difficult aspect of managing small projects is the problem of dealing with many projects at once. This certainly is a problem that the large projects do not have. In the small-project environment, many project engineers and maintenance managers must handle 20 or more projects at once, some of which are in the design and procurement stage, some of which are under construction, and some of which are just being started up. And it must be remembered that many project management activities must be performed regardless of the project's size. The basic problems of small projects are shown in Table 1.1.

As illustrated by Table 1.1, in most organizations we find the paradoxical situation in which small projects, which have the toughest management problems, get the least attention. The reasons for this situation are described below.

Many Small Projects Exist in a Production Environment

Most small projects, involving maintenance, improvements, etc., take place in an operating plant of some sort. This facility operates for one reason: to make a profit by producing the maximum amount of on-spec

TABLE 1.1 Managing Small vs. Large Projects

	50 M$ Project Expenditure	
	Large	Small
No. projects	1	100
No. estimates	1	100
No. schedules	1	100
No. purchase orders and subcontracts	100—200	500—1000
Avg. project duration	3 yrs.	6 months
Full time PM team	YES	NO
Formal PM control procedures	YES	Unlikely

product. Everything about the plant is dedicated to this one goal: it's organization, procedures, priorities, and expertise.

Because the top priority is production, the small projects required to keep the plant running are often considered, at best, a "necessary evil". This typical situation finds the project engineer constantly scrambling for the people, materials, construction equipment, management attention, cash, and even time required to get his projects done. Many of the people on whom he must depend will perform project work only as a parttime, low-priority task. The production environment is also one in which things frequently change, as breakdowns and other unforeseen crises divert attention and manpower to unplanned but highly critical work.

Organization: Not Designed for Projects

As one might expect, the plant organization within which the small projects must be run is generally not designed for project management. The plant organization is, of course, intended to insure production, and must cope with the project work which is required in the best way it can.

Frequently, the project engineer must communicate with, and draw support from, such organizational functions as:

Engineering, drafting
Purchasing, warehouse
Construction, maintenance
Project engineering, planning, estimating
Accounting
Upper management

The project engineer often finds himself in the classic dilemma of having responsibility without the authority to back it up. Since project work is apt to be a parttime, low-priority job for these various support groups, the problem of getting the work done can become acute. And, his problems are often compounded by the fact that he has had little opportunity, at school or work, to be trained in the principles and methods of project management.

Perhaps because of these organizational problems, small projects often suffer from a lack of the formal procedures, methods, and data which are available to larger projects. In spite of the ease with which actual cost and schedule data could be gathered, planning and estimating are often done without the benefit of a good database, and often without experienced planners and estimators. In addition, this problem is frequently made worse by the lack of time available to do the planning and estimating in the first place.

Small Projects: A Special Problem of Control

This lack of a sound plan and estimate naturally leads to problem in project control. Compounding these problems are the special aspects of the small project such as:

Short project life: This leaves little time to gather data, identify problems and correct them.

Shared responsibility: This makes it difficult to obtain commitments and enforce accountability among individuals and departments.

Problems in obtaining actual data: These problems lead to inadequate reporting resulting in a lack of the information required for effective control.

Many projects to be controlled simultaneously: The number of projects adds complexity as some projects will be in early stages of design and procurement while others are under construction. The sheer number of projects handled can often be a problem: Many facilities have hundreds of small projects in progress simultaneously.

Why Standard Approaches Don't Work

Many project engineers in the small-project environment have tried the procedures and techniques developed for large projects in an attempt to improve the management of the smaller projects. These attemts are often unsuccessful because large-project management techniques require a project team with specialists in, for example, planning and cost engineering, and computer applications.

Standard approaches are usually based upon a detailed plan and cost estimate which can be used as a basis for control. Since most small projects do not receive a high enough level of effort in the planning and estimating stage, there is often no basis for exercising project control in the usual way. Also, standard approaches are not suited for the

short timeframe within which small projects are executed, the division
of responsibility, or the problem of dealing with many projects at once.

The fact that most small projects are executed at plant-level is also
a key reason why standard approaches don't work. Plant operations
impose a great many constraints on a small project which do not exist
for large, grass-roots projects. Access is likely to be restricted to the
work area, hot-work permits will probably be required, construction
and maintenance personnel must work among operations personnel
whose work has priority, and the unpredictable nature of plant operations is likely to cause numerous changes to the scheduled access to the
unit and the availability of personnel.

Perhaps the main reason why standard approaches don't work for
small projects is that most small projects are "revamps". What is a revamp project? A revamp project is a change to an existing facility.
Such changes are usually made to improve the unit's performance in
some way. Examples include:

Debottlenecking projects to increase the unit's capacity by replacing
  piping or equipment which is currently limiting performance.
Changes to improve safety, operability, or maintainability by adding
  such items as lighting, platforms and stairways, piping and
  valving, alloy materials, redundant or oversized equipment, etc.
Additional facilities to maintain or improve operations
Facilities to assure compliance with present or anticipated environmental requirements
Modernization projects.

Major maintenance projects often share the special problems of revamps. These special problems include:

Complications caused by working in an operating plant
Congestion in the work area
Lack of access to the work area, which often creates the need to
  perform work in a non-optimum sequence
Interference from normal plant operations
Inadequate design definition causing quantities of piping, steel, and
  electrical materials to overrun because the actual placement and
  routings are not defined until the construction takes place
Engineering manhours often overrun because the scope of work is
  difficult to define
Field manhours often overrun because both the scope and difficulty
  of the work have not been properly defined
Costs often overrun, not only because of increased quantities and
  manhours, but also because overtime or shift work may be required to get the work done when access to the unit is permitted,
  or to get the increased scope of work completed on time.

Each project tends to be thought of as unique, with little in common with previous projects and therefore with no easy way to define it.

Turnaround projects (or "outages"), in which a unit is shut down for major maintenance work which is carried out on a "crash" basis, tend to be better planned and controlled than more routine revamp and maintenance work. Because they usually get attention and support from top management, and because there are no conflicts with an operating plant, turnaround projects are more like large projects for which standard approaches will work well.

Small Projects Need Special Management Techniques

It is certainly evident that small projects have special problems which require special project management techniques. Any project management technique, to be successful in the small project environment, must be able to:

Handle many projects at once
Be used effectively by project personnel with no training or experience in planning, cost estimating, or project control
Cope with the short timeframe of small projects
Simplify the organizational interfaces
Handle the complexities of work in an operating plant
Provide a basis for accumulating cost and schedule data for use in planning and estimating future projects
Improve the project engineer's capability for estimating, planning, scheduling, resource management, expenditure forecasting and control, capture and analysis of project data, and preparation of management reports
Provide a consistent approach to be applied to each project and by each project engineer
Be adaptable to existing procedures

If it were easy to provide such a management system, there would be no need for this book. It is, however, possible to adapt certain proven project management techniques to the special problems of small projects. That is what the remainder of this book is about.

SAMPLE PROJECTS

The Role of the Sample Projects

Because the environment in which small projects are executed varies greatly, the following projects are presented to highlight the special aspects of small projects in various typical environments.

Special Problems in Large Manufacturing Plants

Typical small projects include maintenance, turnarounds, plant ad-
ditions and plant improvements. The problems of the manufacturing
facility center primarily around the overwhelming importance of production and the resulting fluctuations in priorities, as well as the availability of materials, labor, engineering services, and access to the
unit. Projects in manufacturing plants often experience large increases
in the scope of work, as the difficulties involved may not be apparent
until the work is progressed to a certain point. Also, major maintenance projects, such as turnarounds, operate under tremendous time
pressures but the scope of work may be impossible to define until the
unit is shut down and the reactor vessels or other equipment items are
opened up. Many maintenance crews have opened up a pressure ves-
sel to find all it's internals lying peacefully at the bottom.

Special Problems in Large Corporations

There are many small projects that are carried out by large corporations
which do not involve manufacturing, such as engineering and research
work. Although they do not involve construction, these projects share
the same problems of the nonproject environment and multi-project
management. A frequent question in an engineering department in
which 20 engineers are each working on five projects is apt to be, "To
whom should this new work assignment be given?" Engineering projects are also difficult in that physical progress is harder to measure
than in construction work, and that the effect of changes can be far
greater than most people realize.

Research projects are characterized by the very high quality of the
technical, manufacturing and construction work, and the very low (usually zero) profit which results. A research project constantly weighs
the tradeoffs between the technical quality necessary to assure good
results, and the budget limitations.

Special Problems in Small Companies

Small companies are directly or indirectly involved in most small projects. To a small company, of course, a project which is small to the
client may be the biggest one the company has ever handled. In spite
of this difference in perspective, the small-company project is usually
done without the highly-trained personnel or sophisticated systems
which large companies can afford. Small-company projects can also be
critical if they represent a lump-sum contract on which the profit mar-
gin is important. Perhaps no man is more attuned to cost control than
the manager of a "hard-money" contract.

Special Problems of Projects in Remote Locations

Many small projects take place in remote locations such as the Arctic, the desert, the jungle, the mountains, or offshore. Of course, the most important and difficult aspects of these remote projects are logistical; getting men and materials to the site, communications, and difficult working conditions at the site. Resource planning and control is therefore particularly important on this type of project.

CHAPTER SUMMARY

The special problems of small projects are significant and difficult to solve. They are, however, worth solving because of the often-overlooked importance of the projects to the companies involved. Although the specific situations vary, all small projects need proper planning, scheduling, contracting, and management, and the organizational problems can be addressed by techniques which provide a basis for communication, commitment, and control. Fortunately, the methods which are available to manage small projects also provide the necessary basis for resolving the organizational problems as well. These methods are derived from the basic concepts discussed in Chapter 2.

2
Basic Concepts of Small Project Management

INTRODUCTION

In Chapter 1 we discussed the special problems of small projects, explored some of the reasons why standard approaches to project management do not work in the small project environment, and outlined some of the performance criteria we apply to any management tools we propose. In this chapter we define the basic concepts which will then be developed into the specific project-management tools to be described later in this book.

CRITERIA FOR SMALL PROJECT MANAGEMENT TECHNIQUES

From the problem definition provided in Chapter 1, we can see that, in order to handle the special problems associated with small projects, an effective project management technique should have:

1. *A standard approach.* Although each small project is apt to vary significantly from all others, all projects need to be handled with a standard approach. This requirement is to assure efficiency and consistency on the part of all those who are associated with the management of a project (including corporate management). A standard approach is also much more likely to be used and understood. However, a standard approach must be flexible enough to be effective under all circumstances which may be encountered on diverse projects.

2. *Simple systems and techniques.* Although the small project management problem is complex, the diverse nature of the personnel who

will be using (or exposed to) the project-management techniques and the need for ease of training requires that the techniques involved be simple to use and understand.

3. *Fast response.* The short duration of the typical small project indicates a need for techniques which provide information and answers quickly and effectively. This requirement also stems from the rapidly changing environment in which many small projects exist: any change in the operation of the host plant can impact the schedule and resource availability of the small project. This indicates a requirement for flexibility, to be able to rapidly adjust detailed protect plans as the situation changes.

These requirements describe a systematic, consistent, and effective project management and control system which will perform well in the small-project environment. This system can be constructed by using the basic concepts defined below.

BASIC CONCEPTS OF SMALL PROJECT MANAGEMENT

There are four basic concepts which form the basis of an effective small project management technique:

Integration of cost, time, and resources
Use of planning networks
Use of project models
Effective computer applications

To those involved in small projects, some of these basic concepts may seem inappropriate, unecessarily complex, or better suited to large projects. In fact, these basic concepts, which are well suited to any project, are particularly appropriate to the small project environment as they result in significant improvements in management efficiency and effectiveness. This can be seen as each of these basic concepts is discussed in detail below.

Integration of Cost, Time, and Resources

An integrated approach has, in recent years, been demonstrated to be the only truly valid and effective technique for project evaluation and control. By an integrated approach we mean the use of techniques which recognize the dependency of the various parameters of project management. In traditional project-management techniques these parameters are usually dealt with as if they are independent. Integration of project-management parameters is a very powerful concept, worth careful study and consideration. Although developed and proven in large-project applications, the enhanced efficiency and effectiveness which results from the integrated approach has great relevance to small-project applications. Let us begin our study of integrated-project control by examining the reasons for its greatly increased effectiveness.

The Parameters of Project Management

Project management is the process of definition and manipulation of "project parameters" such that the objectives of the project are achieved in an optimum way. In general, it can be said that there are four categories of project parameters:

The *cost* of the project
The *time* it takes to perform the tasks which comprise the project
The *resources* required to complete the tasks
The standards of *quality* which are applied

Even on a small project, these parameters are explicitly or implicitly controlled, and the results of the project—how much it cost, how quickly it was completed, how efficiently the company's resources were utilized, and how well the resulting facilities operated—all depend on how well the task of controlling these parameters was performed. To put it another way, we can say that we are seeking the "optimum blend" of these parameters.

Many engineers overlook the actual objective of a project from corporate management's point of view. From that loftly perspective, it can be easily seen that the "bottom line" as to the success or failure of a given project or group of projects is, in fact, the "bottom line". That is, every project, large or small, exists for one reason alone, and that is to maintain or improve the company's profits. The contribution of an engineering and construction project to company profitability is, of course, determined by the four project parameters listed above. Let us look at each one of these parameters in more detail.

Cost

The cost of a project can be described from several points of view:

The *investment cost*, which is the total investment in the project
Expenditures, which are the cash required each month during the life of the project
Operating costs, required to maintain and operate the facility over it's useful life

As project engineers, we are concerned with controlling the investment cost as well as advising the accounting department on how much cash they can expect to be called on to pay out. Our objective in cost control is, however, not necessarily to minimize the investment cost. It is to insure that the cost implications of all the many decisions and developments that occur are properly considered so that the optimum cost may be achieved. For example, a design change may be proposed in order to improve maintainability and thereby reduce operating cost. The question is: "Is it worth the increased investment cost?" That question can, of course, only be answered if the effect on investment cost is known, and cost control is the means by which we assure that the increased cost is defined in a timely manner such that it becomes part of the decision as to whether or not to proceed with the change. So the optimum investment cost may not necessarily be the minimum cost.

In another typical situation, we might be faced with an unexpected delay in the receipt of materials. Our choices are to let the progress slip, pay a premium to expedite delivery, or accept a later delivery but work some overtime to maintain the scheduled completion date. Clearly, all of these alternatives have some effect on the cost of the project. Once again, our cost-control efforts are directed toward assuring, not that the minimum cost approach is taken, but rather that the impact on investment cost is identified and is incorporated into the decision process.

In both of the above cases we might ask, "Why is the lowest cost alternative not necessarily the best one?" The answer is simply that the best alternative is the one that has the most favorable impact on profitability, and profitability is affected by all four of our project parameters; cost, time, resources, and quality.

Time

Time, like money, materials, or manpower, is a resource which we manage to achieve our project's objectives. Projects vary enormously in the importance which is placed on timing. In some cases, significant improvements in profitability can be attained simply by accelerating the startup date by one day. Or, in the case of projects which are required by legislation, such as environmental improvements, the company's failure to achieve startup by the required date can result in legal penalties which must be avoided by any cost.

We plan and control our utilization of time with the project schedule, in which we identify the points in time at which various project activities will start and finish. Many project engineers and managers view schedules with an attitude of "the sooner the better", and frequently make great efforts to accelerate the start and/or completion of activities based on the assumption that some good must always come out of it. However, the cost implications of such efforts are often overlooked.

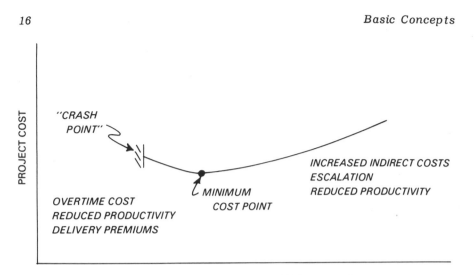

FIGURE 2.1 Relationship between time and cost. Manpower levels are held constant.

The Relationship Between Time and Cost

"Time is money." We all accept that old truism, and it is definitely true for projects. Figure 2.1 shows a curve describing the relationship between time and cost. If we start at the point of minimum cost (i.e., maximum efficiency), and assume that it is desirable to accelerate the schedule we can define a number of ways to do that. Note that, in this discussion of time vs. cost, we are assuming that resources, such as manpower, are held constant.

One possible way to advance the schedule without adding manpower would be to have everyone work overtime. This results in higher costs-per-hour due to the necessity of paying a premium for the overtime hours, as well as the reduced productivity for work done during overtime. This results in more man-hours being expended. In return for these higher costs, some reduction in the time required to do the work can be expected.

Another way in which schedule acceleration can be bought is with the use of incentive payments to suppliers, fabricators, and field subcontractors who are able to beat their schedules. Sometimes these incentives take the form of awarding a contract based on the promised delivery or completion date, rather than the lowest price.

Thus we can see that, in general, a tighter schedule will result in increased costs due to various forms of premium payments and reduced productivity. There is, of course, an absolute minimum time in which the required work can be accomplished. This is often referred to as the "crash point." If would seem only fair that if schedule improvements tend to increase costs, then schedule slippages should result in cost savings. However, the various laws of nature, as well as those of Mr. Murphy, have conspired to assure that this is not the case. What happens to the project's cost when the schedule is extended?

Project costs are often described as having "direct" and "indirect" components. *Direct costs* are those costs which relate to activities which contribute directly to measurable progress. *Indirect costs,* then, are those costs which relate to other activities such as supervision, office services, warehousing, provision of construction equipment, etc. Many indirect costs are determined primarily by time, that is, they are incurred every day or month regardless of the amount of work which is done. Office accommodations and staff costs are good examples: the rent must be paid whether or not the office is busy. It can therefore be seen that, as a project extends past the point of maximum efficiency, the indirect costs will increase.

Productivity can also be affected when the schedule is extended, depending on the reason for the delay. If delays are the result of strikes, work stoppages, or other forms of labor unrest, there will generally be a loss in productivity resulting in more manhours being required to perform the work and an associated increase in cost. Other delays can be due to drawings or materials being delivered late, resulting in work being stopped and started, duplicated, or done out-of-sequence, all of which can result in lost productivity and increased cost. Even the psychological impact of this loss of "momentum" can have a pronounced effect on productivity.

We can conclude, therefore, that cost and schedule are inherently related. Actions taken to control schedule will have some effect on cost. Actions taken to control cost will have some effect on schedule. An integrated approach deals with each of these project parameters as separate terms in the same equation, recognizing that each affects the other and that both affect the project's success.

Resources

Project resources provide the means of accomplishing the work required to achieve the project's objectives, and it is the way in which resources are managed which determines the project's eventual time and cost. In spite of it's importance, this aspect of project management is frequently neglected. To those involved with small projects, resource management may seem to be a lot of extra work to do something which could be done intuitively. However, it is on the small project that resource management is, by far, the most critical since small projects have very little alternative work that can be done when a lack of labor or materials stops progress. Now, we shall look at how resources are related to cost and schedule.

The resources required to accomplish a project generally fall into the following categories:

Professional manpower (e.g., engineers)
Materials (e.g., pumps, piping, steel)
Field labor (e.g., welders)
Subcontracts (e.g., insulation)
Field overheads (e.g., construction equipment)

Our task in project management is to see to it that the required quantity of each resource is available when needed, that it is used properly, and that it is demobilized promptly.

Why is resource management especially critical for small projects? Consider the degree of flexibility which exist on a large project. If an important material item is not delivered on time, if a key piece of construction equipment is suddenly unavailable, or if required crafts are not present in the desired quantity, there is always enough work to be done so that progress can be made in spite of these difficulties. Work in other areas (work with other equipment and crafts) can usually be progressed. But consider the plight of the small project in similar circumstances. There may very well be no other areas to work in, no other material to work on, no way to make progress without the necessary crafts and equipment. And, in addition, no time to waste since the time it takes to sort out these resource problems may represent a major portion of the schedule. So resource management is indeed vital to the small project. To be successful, it must be done efficiently and in a way that recognizes the effect of resources upon project time and cost (see Chapter 4 for details).

The Relationship Between Cost, Time, and Resources

Let us now consider our cost vs. time discussion illustrated in Figure 2.1. What happens if we allow the resources applied to the project to vary as well? Figure 2.2 provides an illustration of the exaggerated effects brought about by resource variations.

Suppose that once again we wish to accelerate the schedule, and now we intend to apply the maximum amount of manpower possible. This will require three shifts working 24 hours per day, seven days a week, which will, in turn, add to costs for shift premiums. Further cost increases will result from reduced productivity due to coordination problems between shifts, lower productivity on evening and night shifts, and increased manpower density. However, these additional extra costs will result in our crash point being moved back even further as we achieve even greater schedule improvement. Should our project be delayed, we can reduce the cost penalties somewhat by reducing resource levels to the minimum which can be used effectively.

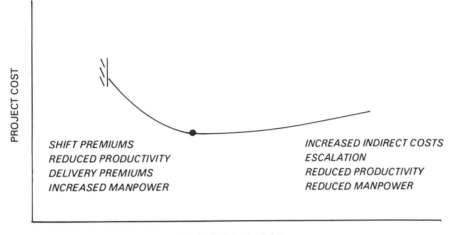

PROJECT DURATION

FIGURE 2.2 Relationship between time, cost, and resources. Manpower levels are varied.

From the above discussion it can be seen that cost, time, and resources must be managed in an integrated way if truly effective control is to be achieved. This approach is particularly appropriate to the small project which can so easily be delayed or overrun when manpower and materials are not available when required. The useful cost-time-resources curve need not be just theoretical: it can be drawn for any project using the technique of "project modelling", discussed later in this chapter. Unfortunately, small projects seldom enjoy the attentions of a large staff of project personnel, so any approach which is to be successful must be handled by the project engineer alone. Fortunately, the integrated approach can produce substantial improvements in individual and organizational efficiency, making the project engineer's job easier and his efforts more effective.

Quality

The technical quality of the work, i.e., the scope of the design which has been specified and the standards to which it will be built, has a clear effect on the cost and schedule. Increased quality usually results in increases in both project cost and schedule length. And, as shown above, the optimum technical quality is that which results in the maximum return on investment, not necessarily the level that costs the most.

Although the importance of quality management is evident, it is a much misunderstood subject discussed further in Chapter 3. For

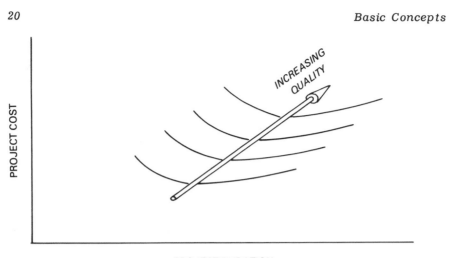

PROJECT DURATION

FIGURE 2.3 Relationship between time, cost, and quality.

example, quality control and quality assurance efforts are usually aimed
at assuring that a certain level of quality is achieved, not that the design quality be optimized for cost, schedule, and return on investment.
Also, quality control should start when the design work begins. The
basic design for a project can be thought of as setting a minimum cost
level that is, given the design, there is a certain minimum project cost
that must be incurred. Though changes are infinitely easier to make at
the early stages of design, how often is there comprehensive design
optimization on a project? In most cases, the design is completed, an
estimate is prepared, the project is built, and the facility operates for
years in what may or may not be the most profitable way.

The relationship betwen quality, cost, and schedule can be seen in
Figure 2.3. As quality is increased, the cost and the time required
are likely to increase as well.

Efficiency Gains from an Integrated Approach

In Chapter 1, the organizational problems associated with small projects
were seen to include a lack of manpower to manage the project most efficiently. The integrated approach is particularly well-suited to small
projects because it avoids the duplication of effort which is often found
in the standard approach to project management.

Let us consider the two major phases through which any project,
large or small, must pass. The first phase might be referred to as the
"project evaluation" phase. In this phase, the basic issues which must
be addressed include:

What will the project cost?
How long will it take to do it?
How will the work be done?
What resources are required to do it?
How much additional profit can be made?

Once the project is approved, it enters the "project execution" phase
in which the design, procurement, and construction take place. During
this phase, efforts are dedicated to the management and control of the
project's cost, schedule, quality, and resource utilization. As we now
know, our true objective is the optimum blend of these project parameters. Both phases consist of efforts by diverse groups involved in
these functions:

Cost estimating
Planning and scheduling
Purchasing and material control
Quality assurance
Construction planning and supervision
Economic analysis
Accounting
Management

In most cases, these functions are performed separately, with meetings, memos, and phone calls being the primary modes of communication. Each function is apt to be performed using informally acquired
data and unofficial methods. This practice can often result in inaccuracy, inconsistency, and inefficiency since there is a considerable
duplication of effort, as well as excess time spent in communication.
All the above functions are similar in that they involve tracking, analysis, forecasting, and reporting of project data, identification of problem areas for attention, and followup of any corrective action which
may have been taken.

An integrated approach, then, provides us with enhanced efficiency
as seen in Figure 2.4. Instead of each organizational function using
it's own data, the integrated approach allows those involved in the projects to use a central database and model of the project.

Use of Planning Networks

The second of our basic concepts is the use of planning networks. Although virtually all large projects these days use planning networks,
they are generally regarded as too cumbersome and complex for small-
project use. There are, in fact, a number of compelling reasons for
using networks on small projects (see also Chapter 3).

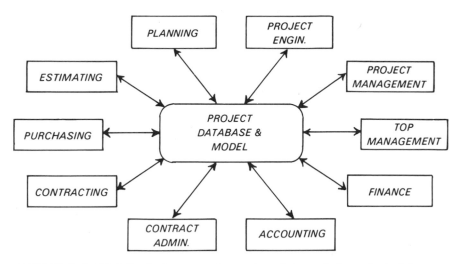

FIGURE 2.4 Distributed processing using integrated project data.

A Network Is the Best Way to Represent a Project

Planning networks have been used for the past several decades, and
have gained much broader acceptance in the past 10 years or so, as
computers became more available for project use. The growing use of
networks is due primarily to the fact that it is the only valid way to
represent a project. The essence of a project is simply many different
people doing many different things over a certain period of time, working toward a common goal. A network allows us to describe those
people and their activities in a rational, quantitative, and mathematical
way. In addition, it is a tool which can be used for many different
applications such as scheduling, estimating, resource analysis, cashflow forecasting, and project control. However, just because networks
sound good on paper doesn't mean they are effective when used on an
engineering and construction project. In fact, many project engineers
and managers who have worked with networks have sworn never to do
so again. In most cases, this is because an excess of planning is generally worse than no planning at all, and, unfortunately, this excess
occurs because the most important rule of planning has been forgotten:
keep it super-simple (the "kiss" rule). As a result, field superintendents and others "on the front lines" have been blessed with computer
printouts that weigh 10 pounds, and contain an indecipherable array
of letters and numbers which seem to have little relevance to the actual
work. So, to be effective for small projects, networks must be kept
simple. If they are, they become an extremely useful tool.

Networks Must be Simple

The human brain is probably incapable of understanding, and using effectively, a network of over 200 activities. In fact, as a general rule,
50 activities is probably a good maximum for a small project. Surpris-
ingly, any project, no matter how large, can be represented by a
simple network: it all depends on how much detail is included in each
activity. We can, in fact, construct a hierarchy of networks, as shown
in Figure 2.5, in which a complete network at one level is represented
by one activity at the next higher level. It also can be seen that the
need for detailed information is, in general, inversely proportional to
one's position in the organization: that is, the higher one's position,
the less detail is required. Thanks to our ability to select a level of
detail for our network which matches the level of detail needed by the
user, networks need never be too complex for effective use (see Chapter 3 for the use of networks).

Network Provides the Basis for Integration

There is, in addition, an important reason for using networks that goes
beyond the general benefits described above: *the network provides
the basis for integration of the cost, time, and resources required to
complete the project.* In other words, the network is our framework
for integrating and manipulating these key project parameters. The
following is a description of how this process works (see Figure 2.6).
For each activity on the network we identify:

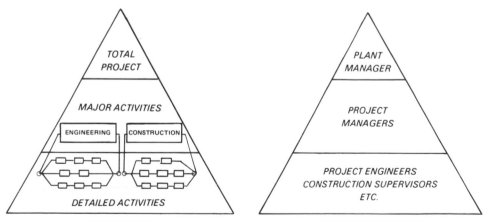

FIGURE 2.5 Heirarchy of network detail.

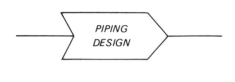

SCOPE: *PREPARE PIPING DESIGN DRAWINGS*

DURATION: *15 DAYS*

RESOURCES: *(1) PIPING DESIGNER*
                 (3) DRAFTSMEN

COST: *(1 DESIGNER) (15 DAYS) (320 $/DAY)*   =   *$ 4,800*
                 (3 DRAFTSMEN) (15 DAYS) (250 $/DAY)   =   *$11,250*
                                                             $16,050

FIGURE 2.6 A resource and cost loaded network activity.

The *scope of work* defined by that activity (i.e., activity description)
The *time* available to perform that work (i.e., activity duration)
The *resources required* (e.g., materials and labor) to perform that
    scope of work in the time available
The *cost* to provide those resources over that period of time

Having defined the cost, time, and resources for each activity such
that they are consistent and in balance, we can say we have a re-
source- and cost-loaded network plan for the project. But why do all
this for a small project? The reason is that the fully integrated project plan becomes the basis for communication, consistency, coordination, commitment, and control. These five aspects of project management tend to be inadequate on small projects.

Communication: This is enhanced when general concepts, plans, and
    intentions are put in specific form and clearly documented.
Consistency: This is improved when each plan, estimate, and prog-
    gress report is prepared using the same approach.
Coordination: This is improved by the ability to see what work needs
    to be done, as well as the consequences of the failure to perform
    that work.
Commitment: This is obtained when an understanding is reached ac-
    according to a specific definition of what is needed, when it is
    needed (and for how long), and a specific agreement is made to
    provide it.
Control: This is possible when, as the project proceeds, an agreed-
    upon plan is available for comparison with what actually happens.

A final advantage of networks is that they provide an excellent means
of computerizing project information. Given the amount of data involved

in small projects, the need for quick results, and the lack of manpower available for data analysis and report preparation, it is evident that effective computer assistance to the project engineer would be most welcome.

If we assume we have a fully resource- and cost-loaded network, and that we have set it up such that we can manipulate the data, we can say that we have created a "model" of the project. Note that, although computer-assisted techniques will be discussed thoroughly in Chapter 11, the methods described in this book do not require computerization when applied to small projects.

Modelling

A project model is a complete mathematical and descriptive representation of the project, expressed in terms of the project parameters of cost, time, and resources. Because it is mathematical, it can be programmed into a computer such that manipulation of the data will show the outcomes of various scenarios.

Why build such a model? To answer that question best, one should consider the better-known models used in industry. For example, we are all familiar with the econometric model used to predict inflation trends, gross national product, and other measures of economic performance, and most engineers have worked with computer models which run calculations to simulate what will happen under certain conditions. Models need not be computerized: consider the wind-tunnel model of an aircraft, used to simulate performance with different configurations, or the model basins used to test ship designs. What these models all have in common is the ability to simulate a real-life situation in order to see what will happen under a given set of conditions. The model is helpful because it would be impractical or impossible to do such a test in real life. For example, government economists don't experiment with the national economy (although there are times when it seems that they are), but they can test the outcome of various decisions by using their economic models. Similarly, it could be quite dangerous to use real aircraft to do configuration testing which can be done safely with wooden models in the wind tunnel.

But why model a project, especially a small one? Because a project is an exercise in the optimization of variables. These variables involve people in real-life situations in which they respond in a very complex way. In addition, a project involves the manipulation of a lot of data with results required quickly. The time-cost-resource curve shown in Figures 2.2 and 2.3 could not have been drawn other than by manipulating the variables with a project model. In fact, a project model is a fancy way of describing something very simple: a way of analyzing the project parameters of cost, time, and resources so we can record what is planned, what has happened, and what will happen. The model is how we accomplish integration.

Effective Computer Applications

The last of the four basic concepts for small-project management is
effective computerization. Much has been said, written, and done
about computer applications, and, since computers currently have much
to offer the project engineer, this book will be no exception (see
Chapter 11). Our task at this point is to examine the question of why
computers are a practical tool for managing small projects. Although
computers should clearly be considered for any small-project application, there are still situations in which manual methods are preferred.
As noted earlier, the basic methods described throughout this book can
be done manually.

Benefits of Computer-Assisted Methods

No doubt most readers will have had extensive experience with computers and seen that they can, in almost any situation:

  Handle a lot of data very quickly
  Provide a standard approach and assure consistency
  Be made to be quite simple for the user
  Assimilate and organize data from diverse sources
  Provide reliable storage of data for use by others.

In addition, we can see that an integrated, modelling approach is
greatly enhanced by computer-assistance.

From our definition of the problems of small projects, in Chapter 1,
it is evident that many small-project problems could be alleviated by an
effective computer system. Yet many companies today are still using
manual techniques even when they are known to be inadequate. One
of the many reasons for this is the fact that, in past years, computerization often turned out to add new problems rather than solve old
ones. The batch-processing mainframe system, which was the only
system available just a few years ago, was often expensive to run, inflexible in operation, slow to return results, difficult to learn, and,
as a result, somewhat impractical.

For many years, information on a printout sheet was automatically, in
the eyes of the reader, draped with the cloak of impeccable credibility.
In later years, as more people became aware of system shortcomings,
the work "garbage" was used more often to describe both input and
output. Today, we have the more reasonable attitude that computer-
generated results may be good or bad, depending on the validity of
the program and the data used.

So there is no doubt that computer systems have something to offer
our hectic, disorganized world of small projects. The questions are:
what can be accomplished, and how can it be done. The answers begin
to be found with recognition of certain trends in computer technology

which have given us, in recent years, some new developments which have profound implications for project management.

Developments in Computer Technology for Project Management

During the past five years there have been developments in both hardware and software that have drastically changed our ability to use computers effectively in the project environment. The most notable advances are:

1. *Distributed processing.* The ability to bring computer power right to the person using it, with workstations from mainframes, or small computers, or both. This provides instant response, with direct input by the user and immediate output from the system, and it means that a computer-based solution to a problem can be implemented quickly and easily. Distributed processing allows all those who might be associated with a project to access and update information in the central project model and database shown in Figure 2.4.

2. *Interactive operation.* The ability to input information or commands via a keyboard and receive an immediate response to the instructions given. In the small-project environment, this means that the time required to capture, analyze, and present data is reduced to the point where one can control projects to an extent far beyond what was accomplished before.

3. *Lower cost.* With the advent of minicomputers, microcomputers, personal computers, and even home computers, the cost of computer-based solutions has dropped to where, in some cases, it is not significant. And, because smaller systems are available off-the-shelf, computer-based solutions can be implemented, if necessary, in a matter of days. This is very helpful in the small-project environment in which little money is available for project-management systems, and often, little time is available for implementation.

4. *Flexible software.* Consistent with the availability of interactive, command-driven systems, software has become available which—instead of providing the long, complex, inflexible programs that the old systems did—allows the user to specify exactly what he wants. This is perfectly suited for small-project management procedures and systems, which have to fit existing company practices that are unlikely to be changed to suit the project engineer's needs.

5. *Project management software.* During the past five years, the need for flexible, effective, and easy to use project-management systems has been recognized by vendors who now offer a very large selection of software and systems specifically designed for project management. As the integrated approach has gained acceptance, the software has also developed to the point where a variety of integrated systems are available, even on microcomputers. This reduces the time and cost

of selecting and implementing a small-project system, and much of this
software is intended specifically for small projects.

   6. *Personal computers*. Low-cost computing capability dedicated to
a single user.

   7. *User-friendly systems*. As more people have become system
users, the systems themselves have become easier to use, i.e., more
"friendly" or easier to learn and more tolerant of mistakes. Since the
small-project environment generally has little time or budget for development and training, and since a variety of people need to have access to the system, user-friendliness is quite important.

Application to Small Projects

It is easy to see how these recent trends in computing are highly relevant to the problems of small projects, and to the basic concepts of integration, networking, and modelling. Although computer-assistance
is not essential in every case to improve the management of small projects, there are probably very few situations in which manual methods
are currently used, that computer-assistance should not be evaluated.
So although the concepts and methods presented here do not have to
be computerized to be effective, the compelling relevance of modern
computer technology to small projects is such that it cannot be ignored
(see Chapter 11).

CHAPTER SUMMARY

We have, in this chapter, defined the basic concepts of project management which will be developed throughout this book for application to the
small-project environment as defined in Chapter 1. Some basic concepts
may seem, at this point, to be too new, old, complex, or simple to be
applied to small projects. While that, in some cases, may be true, it
must be recognized that the basic concepts of integration, use of networks, modelling, and computer-assistance are themselves integrated:
that is, they work best when they are applied together. Each of the
remaining chapters explores an aspect of small-project management,
and shows how these basic concepts are extended into practical tools
for the project engineer.

Part II
Project Planning, Estimating, and Approval

3
Network Planning for Small Projects

PLANNING THE SMALL PROJECT

This chapter describes the use of simple planning networks for the planning, scheduling, estimating, and control of small projects. Although most project engineers operate without networks, we have seen in Chapter 2 that the use of simple networks is very appropriate to the small project. A simple planning network:

Encourages consistency among project engineers

Provide the basis for communication, coordination, commitment, and control

Provides the basis for estimating labor costs, the most difficult aspect of a revamp project

Provides the basis for planning and control of the utilization of material, labor, and construction equipment

Makes possible the improvements in efficiency and effectiveness that accrue from an approach which integrates time, resources, and cost

Is well-suited to simple computer-assisted management techniques

Helps assure that the scope of work is properly defined and understood

Can aid in managing the contractor

For contractors, can be an aid in documenting claims, delays, and extras

Provides a basis for decision-making

Is flexible enough to cope with the changing circumstances which characterize many small projects

Although major improvements are possible from the use of planning networks, it must be emphasized that this technique can be used effectively without requiring special training or complex methods.

This chapter provides a basic discussion of the fundamentals of network planning as they apply to small projects. Although there are many textbooks and training courses which cover this subject in great depth, and are intended for professional planning engineers, the information provided in this chapter will be sufficient for the project engineer involved in small projects. The first part of this chapter deals with definitions. It has been the author's experience that a great many project-related problems are due to a failure to clearly define project-management terminology and to use the terminology in a consistent manner. In spite of the ease with which a solution can be provided, most companies lack these definitions, possibly because the terms used for project management are those which with everyone already feels familiar. It is precisely because of this assumption that everyone will understand every term the same way that problems of communication exist.

Project management is a blend of the diverse disciplines of engineering and management. Practiced mostly by those trained in engineering, it is often the first experience that project engineers have with the ambiguity and subjectivity of management functions. By clearly defining our "variables" as well as the procedures we will apply for dealing with them, we can quantify, as far as possible, the qualitative nature of project management. The definitions offered in this book are not necessarily the only way that these terms can be used. However, they are definitions that have been shown to work well, and they will be used consistently throughout the book. The reader is encouraged to define and document similar definitions to fit his own situation as the ambiguity surrounding one's understanding of these common terms is often immediately apparent when an attempt is made to define them. Although it may appear to be stating the obvious, the definition of terminology is often the most useful first step that should be taken in implementing improvements in small-project management.

WHAT IS PLANNING?

There are a great many expressions regarding planning which are often used loosely, and it is therefore important to start with a clear definition of the terminology which will be used.

"Planning" is the process of breaking a project down into specific tasks, and defining the sequence in which those tasks can or must be performed. Planning is, interestingly, not the same as "scheduling." Scheduling is the process of defining the timeframe in which each task will be performed, and thereby determining the start and completion dates for the project.

We could say that planning defines "what" is to be done, while scheduling defines "when" it is done. The resource plan, which is discussed in Chapter 4, describes "how" the work is done. By clearly defining, before the work starts, just what we have to do, how we intend to do it, and when it will be done, we establish a basic agreement for the small project which is too often lacking in current practice.

DEFINITION OF PLANNING TERMINOLOGY

"Critical Path Method" (CPM): CPM is a planning technique in which a network is prepared and a sequence of network activities is defined such that if the completion of any one of them is delayed, the completion of the project will also be delayed. This sequence of critical activities is called the "critical path." This method is appropriate for small projects and will be used in this book.

"Program Evaluation and Review Technique" (PERT): Although CPM and PERT are often used synonomously, like planning and scheduling, they are, in fact, quite different. PERT is a network planning technique which allows us to deal with the uncertainty of the duration of the activities on our network. Using PERT, we can define a most probable duration, an optimistic duration, and a pessimistic duration for each activity. Statistical approximations are then used to calculate the expected duration for each activity and for the entire project. Because PERT deals with probabilities, it is a technique which is not appropriate for most small projects. For extremely important projects, such as turnarounds, for which the completion date is the paramount concern, PERT may be a useful way to communicate and control the probability of delay. Manual and computerized techniques are simple to apply (see Chapter 5 under Contingency).

"Project": A project is any endeavor for which a purpose, a beginning, and an end can be clearly defined. Those who are involved in small projects know that even this simple definition is sometimes a problem, particularly in defining when the project is complete. It can be seen that, for any project-management effort to be successful, the purpose, the beginning, and, particularly, the end of the project must be clearly understood.

"Activity": An activity is an element of work, a task or series of tasks, which constitutes part of a project. When all the activities are complete, the project is said to be complete. It should be noted that an activity can be defined to any degree of detail, and a project can be defined as having many or few activities depending on the amount of work in each activity. An activity has a duration which is determined by its scope of work, the labor and equipment resources which will be applied, and the basic conditions under which the work will be done. An activity also has an identifiable start and finish. Preferably, it also can be assigned to a specific individual, group, or company.

"Event": An event is a point in the project which has special sig-
nificance. A "milestone" is a commonly used term for event. All projects have at least two events: the start and the finish. Other events,
or milestones, typically include the start of construction, the delivery
of critical equipment, or the award of an important contract. Milestones are a useful way of keeping track of many small projects as they
give a quick indication of whether or not they are on schedule. Unlike an activity, an event (or milestone) has no duration, and no resources are applied to it.

"Network": A network is a pictorial representation of the logical
relationship between the activities in a project. The laws of nature
dictate that some activities, such as preparation of the foundation, must
precede others, such as erection of structural steel. Standard engineering and construction practice also dictates that certain sequences
be followed for the installation of piping, electrical facilities, etc.
These conditions on when an activity can start are "constraints." Many
activities can be done at various times in the project, without inter-
fering with it's overall progress. For these activities, the network
shows the sequence in which we plan to perform the activities, but it
does not necessarily mean that it couldn't be done in another way.
The network simply shows the sequence in which we currently intend
to perform the work. This sequence is the "network logic." Events
are placed in the network to indicate the start or completion of important activities. A network is, fundamentally, a tool for planning.

"Network Notation": Networks can be drawn using two types of
notation:

"Arrow" diagrams: These define the start (or "i" node) as well as
the finish (or "j" node) of each activity. This type of notation is
illustrated in Figure 3.1, and is also referred to as "i-j," "bubble,"
or the "Arrow Diagramming Method" (ADM). An arrow diagram
would quickly become hard to read were it not for the use of "dummy" activities. The dummy simply indicates that two separate
nodes are actually the same: that the two activities, whose finishes
are marked by the nodes at both ends of the dummy, must both be
completed before the next activity can start. By using dummies,
we can draw our arrow diagram such that all activities are hori-
zontal, making it much easier to read.

"Precedence" diagrams: These define each activity in terms of the
activities which must be completed before that activity can start.
A precedence network is shown in Figure 3.2. It can be seen that
a precedence network does not use nodes or dummies. Although
the illustration shows the common finish-to-start relationship,
it is also possible to diagram start-to-start and finish-to-finish
relationships. Precedence notation is also called "Precedence Diagramming Method" (PDM). The activities which preceed the activity

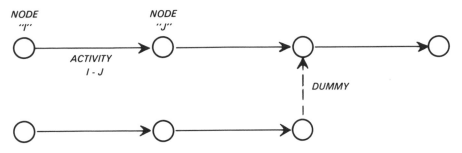

FIGURE 3.3 Arrow Diagramming Method (ADM). Each activity is defined by its start node 'I' and its finish node 'J.'

in question are called "predecessors," and those which come after it
are "successors".

In comparing arrow and precedence notation, we generally find that
precedence is easier to understand and use for those who are not
trained planners. However, since one need only specify two node
numbers to identify any activity, no matter how many predecessors it
has, arrow format can be simpler for large, complex networks. Most
computer programs available today can use either notation, although
most are written initially for one and then adapted to the other. Precedence diagramming has the additional advantage of being able to
describe the start of an activity after its predecessor is only partially
complete. This is done by using a start-to-start relationship plus a
delay in the start of the succeeding activity. In ADM notation, the

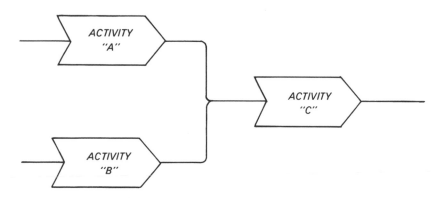

FIGURE 3.2 Precedence Diagramming Method (PDM). Each activity is
defined by specifying those activities which must precede it (e.g.,
activity C cannot start until activities A and B are complete).

preceding activity would have to be split, in order to describe the same relationship.

"Critical Path": After a network has been drawn—showing the activities, their durations, and constraints—it is evident that there is a certain sequence of activities (or "path through the network") whose total duration exceeds that of all other paths. That is, the sum of the durations of each of the activities on this path will be equal to the total time required to complete the project. *If any activity on this path should slip, the completion date of the project will slip a like amount.* This path through the network is the "critical path," and all the activities on that path are "critical activities." Critical path is a very useful concept for small projects.

"Float": For those activities not on the critical path, it is evident that some slippage in either the start or completion can be tolerated without causing any slippage in project completion. This amount of slippage is referred to as float (also called "slack"). There are two types of float:

> *"Total float"* (*TF*): This is the amount of time that completion of an activity can slip without affecting the completion date of the project. Total float is the best understood and most commonly used float calculation. The word float denotes total float.
>
> *"Free float"* (*FF*): This is the amount of time that completion of an activity can slip without it affecting another activity. For example, an activity can slip by less time than its TF, and still affect the starting date of its successor. Free float, although it is a useful control parameter, is seldom used as a practical project-management device, as it requires a good comprehension of network analysis. This is unfortunate, as FF provides the best indication of the amount of slippage that can be easily tolerated (See "Misconception: Float is a Form of Contingency," later in this chapter). It is also useful to indicate which activities are likely to become critical.

Activities on the critical path have, of course, no float.

"Time Analysis": Float times, and the critical path, are calculated by a procedure known as time analysis. Time analysis consists of a "forward pass" and a "backward pass" to determine critical path, float times and the "early-start," "late-start," "early-finish," and "late-finish" times for each activity.

> *"Forward pass"*: Beginning at the "start" of the network, durations are added to determine the early-start and early-finish times for each activity.
>
> *"Backward pass"*: Beginning at the "finish" of the network, durations are subtracted to determine the late-start and late-finish times for each activity.

"Early start time" *(ES)*: This is the earliest time at which an activity can begin. The early-start time will be identical to the early-finish time of the preceeding activity.

"Early-finish time" *(EF)*: EF is the earliest time at which an activity can be completed.

"Late-start time" *(LS)*: This is the latest time at which an activity can begin without causing a delay in project completion.

"Late-finish time" *(LF)*: LF is the latest time at which an activity can be completed without causing a delay in project completion.

From the above it can be seen that, for any given activity:

$$TF = LS - ES = LF - EF$$

Where

$ES = LS$, $EF = LF$, $TF = 0$, the activity is critical.

"Hammock": In order to simplify a network for those with no need for details, a hammock is often used. A hammock, illustrated in Figure 3.3, allows us to represent several activities in the network with one activity. The concept of the hammock is very important to the small project in which we need to avoid the use of detailed reporting, while preserving the level of detail necessary for effective control. Hammocks make it possible to display a network at various levels of detail, to suit the point of view of the user. A management network, often referred to as a "Level 1," provides the summary-level view a

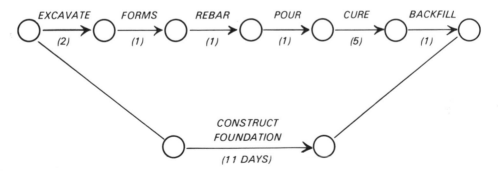

FIGURE 3.3 A network hammock. The single hammock activity, "construct foundation," represents the scope of work and total duration of the six detailed activities.

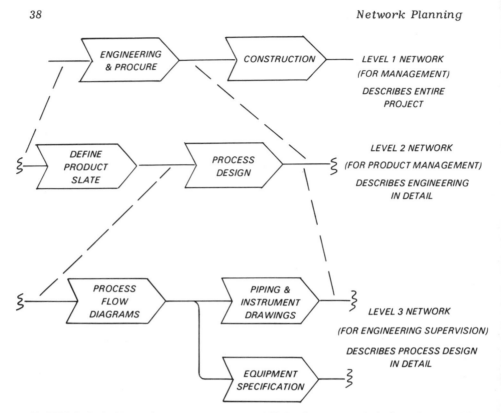

FIGURE 3.4 Use of hammocks to establish the network heirarchy. This example shows the activities for an engineering project.

manager requires. By contrast, the drawing-office supervisor might be working with a detailed plan which shows all the activities associated with each drawing.

This "network heirarchy," made possible by hammocking, is also a useful tool for managing the contractor or subcontractor. For this purpose, a "subnetwork" (or "subnet") is drawn to represent the work described by the single activity on the higher-level network. The subnet allows detailed control procedures to be implemented without causing that extra detail to be visible at the management level. The use of subnets is shown in Figure 3.4.

"Barchart": A barchart is a tool for scheduling, that is, for representing the dates during which each activity will take place. Unlike a network, a barchart does not show the constraints between activities.

FUNDAMENTALS OF CPM

The techniques for planning and scheduling a small project using the Critical Path Method are described in this section. The basic technique consists of the following simple steps:

1. Define the scope of work of the project
2. Break the project down into activities
3. State the assumptions on which the plan and schedule will be based
4. Assign an expected duration to each activity
5. Define activity constraints and draw network
6. Calculate critical path and float
7. Schedule activities
8. Optimize plan and schedule by iterations of steps 3—7

Step 1: Define the Scope of Work of the Project

As most contractors know, the failure to clearly define and agree on the scope of work prior to starting a project is one of the most frequent causes of cost overruns and schedule delays. There are a number of reasons for this, such as the common perception that time is short and everyone understands what has to be done anyway. Another is the feeling that, if the full scope of work is revealed, the project might not be approved. Even when design specifications are prepared in detail, experience shows that contractors will often encounter many unforeseen problems on a small project. All of this often is due to the inherent difficulty in defining the scope of the small project due to its revamp nature.

In preparing a plan for the small projects, we generally being with a technical description of the project: marked-up drawings, design specifications, etc. Our first step in defining our scope of work for planning purposes is, therefore, to assure ourselves that the design basis for our plan is reasonably complete. This might entail a trip to the field (to clarify both the design and construction work required), or a review of similar past projects. We should be able to reach agreement on the documented technical scope. Given an agreed and understood technical scope, are we now ready to being planning? Unfortunately, we are not. Like so many other aspects of small projects, a special problem arises: the problem of "who does what." The point at which the project is handed over to the project engineer and the point at which he completes it and hands it back to his client will vary and must, therefore, be defined. Many project engineers have been heard to cry, "But I was told that the site clearance would already be done before we arrived!", or, "They said that all the buildings and utilities for construction would be provided!" Of course, when these things have not been done, we are left to do them ourselves and hope we can get a scope change approved later on.

When is a project "complete"? Does a project's scope include startup and commissioning of all systems? Does it include final revisions to all drawings? Does it include all "punchlist" items? The answer is, "it all depends." Some projects include these items, and some don't.

If a clear definition of the technical and planning scope of work cannot be easily obtained, the project engineer should clearly document his assumptions, and thereby set the basis for discussing changes if the assumptions turn out to be incorrect.

Step 2: Break the Project Down into Activities

With the scope of work clearly defined, we are now able to define the specific activities required to accomplish that work. As noted previously, an activity can be defined to any level of detail, depending on the use to which the network will be put. We can say that the project engineer's level of detail is Level 2; that is, more detailed than that used by a manager but less detailed than that used by those who are directly responsible for only one part of the project.

What determines the level of detail of an activity? The level of detail should be appropriate to the amount of planning and control required. For example, we might have a project in which the engineering work is to be done on a "cost-plus" basis. Clearly, this is a case in which close monitoring of manhours and progress will be required. In this case, our project engineer's network will contain all the activities describing major aspects of the work. Suppose that same engineering work was to be done under a lump-sum contract? Although he is still responsible for the contractor's performance, our project engineer's network will probably show mostly those activities relating to technical quality, as no control of manhours is required.

The activities should be consistent with the contracting plan, and defined in sufficient detail to provide a basis for resource planning, cost estimating, progress measurement, and control. Most small projects can accomplish this with networks of 50 activities or less. Each activity should be significant, cover an identifiable scope of work, and have a clearly defined start and finish. In the case where the detail required for control exceeds the detail of the Level 2 network, we can use a subnetwork to describe activities for control as shown on Figure 3.4. Some companies have found that a planning checklist is helpful to assure that activities, such as testing, which can be overlooked but are nonetheless important are accounted for in the network.

Step 3: State the Assumptions on Which the Plan and Schedule are Based

In spite of all the attention given to network planning techniques, a plan or schedule is only as good as the assumptions on which it is based. Some of the assumptions which have a marked influence are:

   Work schedule
      number of shifts per day
      number of working days per week
      use of overtime
      effect of holidays
   Work content
      conditions at start
      definition of project completion
      amount of work which is currently undefined
   Quality requirements
      normal practice
      above-normal practice
      sacrifice of quality for work speed?
   Cost vs. schedule priorities
      OK to spend money to expedite schedule?
      is minimum cost the primary objective?
   Availability of resources
      skilled labor
      critical materials
      special construction equipment
   Efficiency of management decision-making
      can critical decisions be made quickly?
   Access to work site
   Contractor performance
      productivity
      mobilization time
      quickness of response
   Market conditions
   Labor conditions
   Supplier performance
      time to bid
      delivery time
      quality vs. the need to fix in field
   Weather conditions

Assumptions should be made to reflect the most probable situation, not, as is so often the case, the most optimistic. When a database or file of similar past projects is available, the assumptions should reflect consideration of how the project at hand could be different from the average past project. Even if the assumptions prove to be wrong—in fact, especially if they are wrong—the fact that they were clearly

documented in the initial plan makes the project engineer's position a
lot more tenable. If one or more key assumptions are open to question,
a plan and schedule can be prepared for each case, so that the sensitivity of the schedule can be seen.

Step 4: Assign an Expected Duration to Each Activity

The expected duration for an activity is that elapsed time which we
expect will be required to complete the scope of work described by the
activity, with the resources we expect to be available, if our basic assumptions are correct. By definition, the probability is 50% that the
actual duration will be equal to or less than the expected duration (see
Figure 3.5).

   This definition, given to us by statistical methods and illustrated by
PERT theory, points out an important concept which is often overlooked, even by experienced planners. If we calculate the most-likely
activity duration, that is, the duration that will occur more frequently
than any other duration, we find that the probability of that duration
actually being achieved is generally less than 50%. This is due to the
fact that the most optimistic duration is, typically, around 70% of the
most likely (you can only accelerate the schedule just so much), whereas the most pessimistic duration can be 300% or more of the most likely.
This illustrates the principle of "skewness," and is as good an illustration of Murphy's Law as we are likely to find. Because, in general,
there is a limit on how much we can improve things, while there often

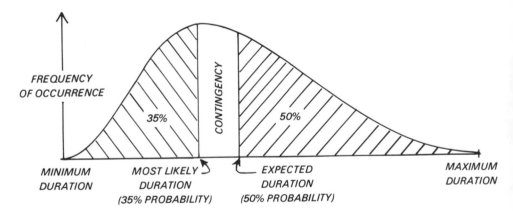

FIGURE 3.5 Probability function for activity durations. The area
under the curve indicates the probability that the actual duration will
be equal to, or less than, a given duration. The difference between
the expected and most likely durations can be considered to be schedule
contingency.

seems to be no limit at all of how much things can be wrong, the expected (or 50%-probability) duration, will be larger than the duration we say is most likely. The consequences of confusing the most likely and expected values will be explored later in this chapter (see "Some Popular Misconceptions," and also in Chapter 10.

The best method for establishing the expected duration is the use of historical data. Most small projects are similar to other small projects by the same company, and it is therefore possible to develop data records which show, on average, how long certain tasks take. If such a database is available (see Chapter 6 for details on how such databases can be constructed), we need only examine the project at hand for differences with the average past project, and make adjustments accordingly. If such information is not available in-house, there are a number of industry services which provide such data. Before using such a service, the database should be checked against some actual data, and factors defined to be applied to the results from the service. Finally, there are numerous judgements and rules of thumb which can be used. The network can, of course, be used to test the sensitivity of the schedule to the durations assigned to various activities.

Step 5: Define Activity Constraints and Draw Methods

Now that we know the activities which constitute the project we are ready to plan the sequence in which those activities will be performed, and present that plan in the form of a network. To establish our network logic, we first must identify those activities which have constraints; that is, those activities which cannot start until one or more other activities are complete. We can group our constraints into two categories:

1. Mandatory constraints. Those physical or operational constraints which cannot be violated; for example, formwork and reinforcing steel must be installed before concrete can be poured. These constraints are sometimes called "hard logic" constraints.

2. Arbitrary constraints. These constraints represent good practice and reflect the way in which we intend to do the work, but can be changed. For example, we could intend to complete a certain amount of engineering work in-house before engaging a design contractor, but, if our drawing office gets overloaded, we may move the contractor's involvement forward. Or though a pump should be installed before its piping, if delivery of the pump is delayed, the piping can be installed first and final corrections made later.

As the constraints for each activity are identified (in terms of those activities which precede it), it is helpful to note which category the constraints fall into, so that the logic can be changed, if necessary, to optimize the network. The network should be drawn, assuming that no constraints are imposed by resource limitations.

A network can now be drawn to display the logic. When the network is complete, it is useful to check for "loops," i.e., errors in logic which create a circular flow instead of a left-to-right linear flow. Another potential error to check for is "dangles," i.e., errors in logic which leave an activity with no connection to another activity.

When drawing the network it is also useful to identify the milestones which will be used later as a quick reference on schedule status. Some companies have found it helpful to define standard networks for groups of activities or for entire projects which are often repeated. Maintenance projects, some engineering projects, and certain types of light construction lend themselves to this approach. Standard networks reduce planning time, increase consistency, and help assure that all necessary activities are included in the plan. The network plan for a specific project can be derived simply by changing and adding activities or constraints as necessary. It should be noted that the network can be drawn freehand on ordinary paper: expensive graphics services are not generally necessary for the small project.

Step 6: Calculating Critical Path and Float

In this step, a time analysis is performed on the network using the following procedure.

Starting at the "start" event, a *forward pass* is made. For each activity, in turn, we write down the early-start time using the notation shown in Figure 3.6. For the first activity (or activities) the earliest possible time that the activity can be started is at an elapsed time of zero. If the duration of the first activity is 5 days, then the earliest time at which that activity can be completed is at an elapsed time of 5 days. This is also the earliest time that the succeeding activity can start. The forward pass procedes through the network in this manner. It can be seen from Figure 3.7a that the early-start time of an activity with several predecessors will be the latest of the early-finish times among the preceeding activities. When the finish event is reached, the network should show the elapsed time to each activity's early-start and early-finish.

In order to calculate the critical path and the float of the noncritical activities, a *backward pass* is made. Starting at the finish event, the latest finish of the last activity can be seen in Figure 3.7b to be equal to the early finish. If more than one activity precedes a finish event, the latest finish will be equal to the longest of the early-finish times. By subtracting the duration of the last activity from the late-finish time, the latest-possible start time is identified. The backward pass continues in this manner. When an activity has several predecessors, its late-start time becomes the late finish of all the predecessors.

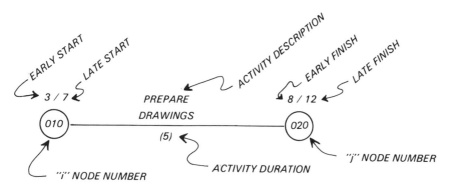

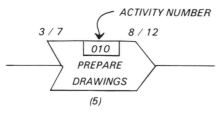

ADM

(a)

ACTIVITY NUMBER

PDM

(b)

FIGURE 3.6 Network notations: (a) ADM. (b) PDM.

When several activities have the same predecessor, the lowest late-start time becomes the late-finish time of the predecessor, as shown in Figure 3.8. When the backward pass is complete, the late-start and late-finish times are shown for each activity.

The critical path is now evident from inspection of the network. Those activities for which the early start/late start and early finish/late finish times are identical are critical. There should be a path through the network made up of critical activities; if not, there is an error.

For those activities for which there is a difference between early start/late start and early finish/late finish, that difference is equal to the total float.

It should be noted that there may be more than one critical path.

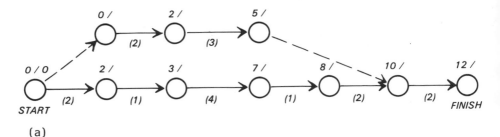

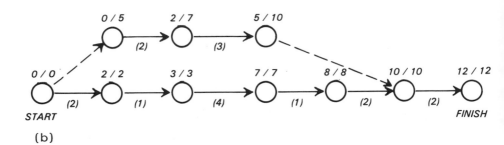

FIGURE 3.7 (a) Forward pass: calculating early-start and early-finish times. (b) Backward pass: calculating late-start and late-finish times. The backward pass reveals that the critical path is the lower sequence of activities and that there is 5 days total float for the upper sequence of activities.

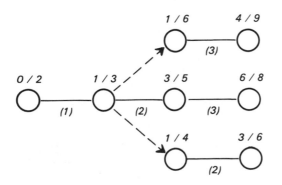

FIGURE 3.8 Backward pass: a special case.

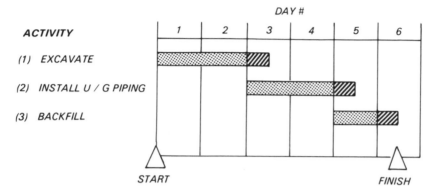

FIGURE 3.9 Bar chart showing total float. Each activity in the sequence has one-half day total float. The bar chart shows early start for all activities.

Step 7: Schedule Activities

With the time analysis complete, the project is now ready for scheduling, that is, defining the date or time at which each activity will begin and end. Given a start date, the dates for early/late start/finish can be easily determined, and the project-completion date defined.

A barchart is the most common way of representing the schedule. For those activities with float, the barchart can show them scheduled to start at the early-start date (leaving the float unused), the late-start date (all float used prior to the start), or some time in between. Barcharts often indicate the float available by using a different shade for the float time (see Figure 3.9). It is important to distinguish between the uses of barcharts and networks.

Networks	Barcharts
Are used for planning	Are used for scheduling
Show constraints	Do not show constraints
Show durations and elapsed time	Show durations and dates
Change infrequently during project	Change frequently

A barchart should always be based on network logic, even if that logic is not shown. Many companies have found that a format combining the best features of the barchart and the network can be useful. Some of these formats are:

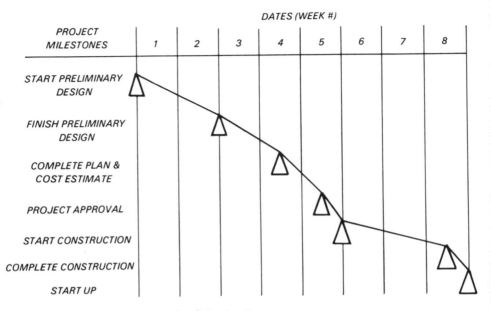

FIGURE 3.10 Sample waterfall chart.

A network plotted on a time scale
A barchart showing constraints
"Waterfall" or "cascade" chart showing milestones and dates

An example of the waterfall chart is shown in Figure 3.10.

Step 8: Optimize the Plan and Schedule

It is evident from the procedures described above that planning is an
iterative process. The assumptions on which the plan and schedule are
based are made, in general, without a good idea of the actual impact
they will have on the project. For example, it may have been assumed
that certain activities will be performed on a crash basis—until the time
analysis reveals that those activities have sufficient float to make that
unnecessary. Or, our logic diagram may have resulted in a critical
path which included activities which need not be critical, so a simple
change in the logic is all that is required to improve the schedule. The
scheduled completion date is often found to be unacceptably late. The
network logic and the basic assumptions must then be examined and
changes made accordingly. As we change start dates, logic, or durations—and iterate in order to optimize the schedule—we are, in effect,
using our project model for decision-making.

Planning and scheduling, in spite of all the arithmetic involved, is still something of an art. Much of the artistry takes place during the iterations of the network, as the variables are changed until a satisfactory plan and schedule are achieved.

The work involved in preparing a network plan and barchart schedule for a small project is generally simple enough that it is quite practical to do manually. However, there are now available many computer-assisted approaches which are both practical and cost effective for small projects. By greatly reducing the time and effort required, these systems encourage the user to make the best use of the network-planning technique, by using it for decision making, "what if?" studies, resource planning, cost estimating, expenditure forecasting and project control. These subjects will be covered in detail in later chapters.

SOME POPULAR MISCONCEPTIONS

As network planning has gained in use over the past decade, a number of misconceptions have grown around it. For the technique to be effective in the small-project environment, these misconceptions must be avoided.

Misconception: Float Is a Form of Contingency

Because TF shows the time that an activity can slip without affecting the end date, it is often thought that it makes no difference whether the float is "used" or not. The float is therefore thought to be a form of contingency, a "slush-fund" of time to take care of unforeseen problems. This is a dangerous misconception that can lead to large schedule slippage.

Let us consider a series of activities each of which has float (see Figure 3.11). Our barchart at the start of the project shows each of the activities with a TF of $\frac{1}{2}$ day. Suppose that the project is now underway, and the first activity has been completed at the late-finish date, thereby using all of it's TF. It is immediately apparent that the remaining activities in the series must all now start on their late-start dates. They have, therefore, lost all their float. So, float is too easily lost to be considered contingency. Another reason that float cannot be considered contingency is that it provides the flexibility needed to compensate for imperfections in the network logic.

Although the concept of FF is seldom used in the small-project environment, it is, in fact, a truer representation of the amount of slippage an activity can tolerate before it affects the early start of another activity. In other words, the FF can be used up without affecting any other activity. An example is shown in Figure 3.12. Free float is the

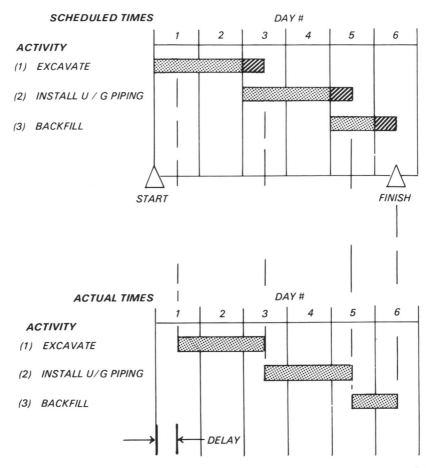

FIGURE 3.11 Effect of losing total float. If excavation begins of its late-start time, thereby "using" its float, it and all succeeding activities become critical.

difference between the early finish of an activity and the earliest of the early-start times for its successors. It is particularly useful when someone else has the responsibility for the succeeding activity. When we use our TF, he is affected as his float is lost, but when we use our FF, he doesn't know the difference.

Misconception: Target Schedules Help Assure On-Time Completion

Many project engineers and managers believe that a tight schedule helps to put pressure on those doing the work, thereby helping to assure

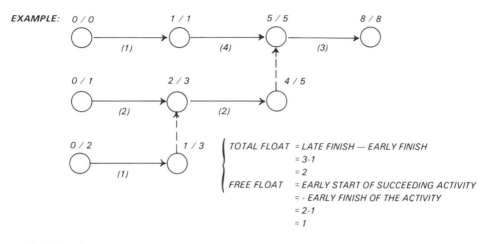

FIGURE 3.12 Two types of float. Total float affects the end date and other activities while free float is the slippage in an activity which can occur before any other activity is affected.

on-schedule completion. One reason why this seldom works is the fact that this approach has been used for so long that it hardly fools anyone anymore. Most contractors know that the client is apt to have a budget and schedule that contains some contingency, and to conceal this fact only leads to a loss of credibility (see Chapters 5 and 10 for further discussion of contingency). Another reason is due to a simple but often overlooked principle of statistics. Most of us remember high-school mathematics and the fundamental principle that the probability of two things happening is the product of their individual probabilities. For example; if the probability that a fair coin will turn up heads is 0.5, the probability that two successive flips will yield two heads is $0.5 \times 0.5 = 0.25$. If we have a target or "tight" schedule, we are saying that the probability of the planned duration of each activity actually being achieved is something less than 0.5—typically it might be 0.35. Now, suppose the critical path contains a node which joins three critical activities. The probability that the next activity will start on time is 0.35 raised to the 3rd power or .04! No wonder these tight-target schedules are so seldom achieved.

To avoid this problem, the project engineer should be sure that the durations assigned to each activity have an equal probability of being exceeded or underrun.

Misconception: A Project With A Tight Schedule Has No Time For Planning

This misconception is widely held in large and small projects alike. It is invariably advanced by those who have not experienced a well planned and controlled project, those whose experience with formal planning techniques has not been good, and those who have just been lucky.

The project-planning function on large projects often suffers from a lack of realism and timeliness, which greatly reduces its credibility and effectiveness. On small projects, there is often no planning at all, due to a lack of experience, methods, data, and training. Once planning techniques appropriate to small projects have been implemented, most companies find the effort expended to plan the work properly is paid back many times in improved communication and control.

Simplicity is the key to successful implementation of proper planning and management of small projects; as long as the methods are simple to use, and appropriate to small-project application, they will be effective and accepted.

Misconception: The Network and Barchart Form a Complete Project Plan

Although the complete network and barchart would seem to be more than enough to assure that the schedule is realistic, there are two important factors which are often overlooked: 1 the plan for the application of material and labor resources, and 2 the contracting approach which will be taken to assure that those resources are provided in a cost-effective way.

The resources which are applied to the project determine, more than any other single factor, how long the project will take. Work cannot be accomplished without the people, the machinery, and the materials necessary to do it. Yet, in spite of the importance of resource analysis to identify resource requirements and assure that the manpower, machinery, and materials are available when needed, this technique is seldom used. Resource analysis is particularly vital for small projects because of the small project's lack of alternative ways to make progress if essential resources are not available when required. For example, if a small project involves replacement of a heat exchanger, and that heat exchanger is not delivered on time, that project will certainly be delayed. By comparison, a large project will have other materials on-site, so progress can be made. Because of its importance to small projects, resource analysis will be described fully in Chapter 4.

The contracting plan describes how contracts will be arranged to assure timely provision of the resources and services required. Many project engineers overlook the extent to which contracting options are available, and the extent to which the selection of a contracting approach can affect the schedule. This is discussed in the next section.

CONTRACT PLANNING

In many small projects, much of the work is done by contractors or subcontractors. The approach taken for bidding, negotiating, and managing these contracts can have a major impact on the project's cost and schedule, as well as on the quality of the work done. Most small projects are conducted without proper attention to defining a contracting strategy which is right for the project. The optimum blend (referred to in Chapter 2) of cost, schedule, and quality will be different for different projects, and the contract plan should be designed to reflect these basic needs.

The contract plan for a small project may be nothing more than a thought process which considers the special needs of the project at hand and the different contracting approaches that are available, and makes a decision on that basis. The network plan and schedule can also be used to evaluate different contract-plans scenarios. For example, one plan might cost more but result in some schedule improvement, and the network model could be used to quantify the schedule benefits that could be expected.

In this and subsequent sections we will refer to the relationship between "client" and "contractor." It should be recognized that every organization is both a client and a contractor: the project engineer is, in effect, a contractor to the operating plant which is often referred to as his client. The engineering and construction contractor is, of course, the client to the firms to which he issues subcontracts. So it is important to consider the client/contractor relationship with an appreciation of both points of view.

Factors Which Determine the Contracting Plan

Some of the factors which affect the selection of a contracting plan are:

Market conditions: Are the contractors looking for work or do they have all they can handle?

Project size: Is this project bigger or smaller than most other projects the contractor has done before?

Contractor capability: Has the contractor demonstrated the ability to handle the size, complexity, or technical discipline of the type of work involved?

Schedule: Is the schedule particularly important on this project? Is time available to pursue a contracting approach which requires more time for negotiation? Is cost more important than schedule?

Technical security: Does the project involve proprietary technology (many research projects fall into this category)?

Financing: Are there any special financing considerations? Is the contractor expected to do the work and be paid later, thereby financing the project himself? Should he be? Can money be saved if the client finances the project at the lower interest rates he can probably secure?

Risk: Are there any special technical, cost, or schedule risks associated with the project? For example, are the penalties for late completion greater than normal? What is the impact of a cost overrun? Is there any uncertainty about the technical basis, i.e., has the design ever been built before?

Company image: Are there any limitations on the selection of contractors imposed by the desire to project a certain image? For example, many companies prefer to work with local or domestic contractors and suppliers to foster goodwill, even when better price and performance could be obtained elsewhere.

The answers to these questions will be different for most projects, and the contracting plan should be designed accordingly.

Basic Types of Contracts

There are three types of contracts: lump-sum contracts, reimbursable contracts, and contracts that are combinations thereof. These are summarized in Figure 3.13.

Lump-Sum Contracts

Lump-sum contracts (also referred to as "fixed-price" or "hard-money" contracts) set a price for a defined scope of work. A pricing structure is also agreed upon for adjusting the lump sum for changes made by the client. The lump sum includes the contractor's direct costs, overhead, and fee. The lump-sum price may be paid in full upon completion of the work, but it is most often paid in increments according to progress.

Advantages of lump sum

The theoretical advantages of the lump-sum contract are:

Competitive pricing: In market conditions characterized by good competition, the lump-sum approach can result in the lowest-possible price being obtained, as contractors may be motivated to cut profit margins while maximizing performance. With a lump-sum contract, the contractor has an incentive to perform in a cost-effective manner.

Reduced risk to the client: Because the final cost is known in advance, the risk of cost overrun is theoretically eliminated. Any overruns should be to the contractor's account.

ATTRIBUTE	LUMP SUM	REIMBURSABLE	REIMBURSABLE W/FIXED FEE	UNIT PRICE
PRICING	HIGHLY COMPETITIVE	COMPETITIVE	COMPETITIVE	COMPETITIVE
CONTRACTING TIME REQUIRED	LONG	SHORT	SHORT	MEDIUM
SCOPE OF WORK DEFINITION	DETAILED DEFINITION, FIXED SCOPE	GENERAL DEFINITION, VARIABLE SCOPE	GENERAL DEFINITION, VARIABLE SCOPE	SEMI-DETAILED DEFINITION, VARIABLE SCOPE
CLIENT RISK OF COST OVERRUN	LOWER	HIGHER	MODERATE	MODERATE
POTENTIAL FOR CLAIMS	HIGH	LOW	LOW	MODERATE
MARKET CONDITIONS REQUIRED	COMPETITIVE	NONE	NONE	MODERATELY COMPETITIVE
NEGOTIATION EFFORT	HIGH	LOW	MODERATE	MODERATE
CONTROL AND ADMINISTRATIVE EFFORT	LOW	HIGH	MODERATE/HIGH	MODERATE

FIGURE 3.13   Summary of contract types.

Reduced control required by the client: Since the price is fixed, no cost control is necessary

Because of these attractive advantages, many project engineers and managers believe that contracts should be lump sum whenever possible. This is a very dangerous policy which has resulted in a lot of unnecessary overruns and many unpleasant negotiations. In fact, contracts should be lump sum whenever it fits the project's design and schedule requirements and the prevailing market conditions. There are, however, some serious disadvantages to lump-sum contracting.

Disadvantages of lump sum

The disadvantages of the lump-sum contract should be carefully considered before a decision is made:

1. The scope of work must be well defined. If any of the lump-sum advantages are to be realized, the scope of work and price which is initially agreed upon cannot be significantly changed. When changes to the scope do occur, the contractor is in a noncompetitive situation and can be counted on to take full advantage of it. This is a particular problem for many types of small projects, such as revamps, turnarounds, and major maintenance in which the scope of work cannot be known with any certainty until the project is well along.

2. The schedule must provide adequate time for the client to define the scope of work to a sufficient level of detail that it can be the basis of a firm bid with low probability of significant changes.

3. The schedule must provide sufficient time for the contractors to prepare bids, and for the client's bid review and selection process. On small projects, this amount of time may be a disproportionate part of the overall schedule.

4. The client's risk is only partially reduced, due to the understandable reluctance of the contractor to let one contract put him out of business. Should changes, unforeseen circumstance, or client-contractor communication problems reach a high enough level the contractor can, quite justifiably, be expected to submit a claim. Once the work is underway, the contractor has the upper hand in any negotiation because the cost to the client of the contractor's failure to perform probably far exceeds the value of the contract. In situations where there is significant risk, the contractor's bids may include contingencies to such an extent that any cost savings from the lump-sum approach are lost.

5. The client is still obliged to monitor and control the progress and technical quality of the work, as well as be alert for potential claims. Since the price has been fixed, quality control is especially important to assure that no corners have been cut to save cost.

6. The willingness of bidders to invest the time and effort required to prepare a lump-sum is very much a function of market conditions. If the contractors are busy on profitable work, they may either decline the opportunity to bid or simply submit a high bid. So, successful lump-sum contracting requires favorable market conditions in which several qualified bidders are willing and able to prepare bids.

In summary, many companies have made the mistake of rushing into a lump-sum contract with an incompletely defined scope of work, only to find that the subsequent scope changes resulted in substantial cost overruns and schedule delays, as well as costly negotiations, all of which could have been avoided. Lump-sum contracts are indeed a way to save money and extract the best performance, but these results only occur after careful planning and execution of a lump-sum contracting plan under the right conditions.

Reimbursable Contracts

Reimbursable contracts (also called "cost-plus" or "time and material" contracts) are those that reimburse all of the contractor's direct costs fully, in addition to which he is paid for his overhead and fee. In its most common form, this type of contract provides an all-in rate for each hour of services provided, as well as a markup on materials and services procured on behalf of the client.

Advantages of reimbursable contracts

1. Shorter contracting time. Because it is only necessary to define the scope of work for a reimbursable contract in general terms, the time required for the client to define the scope and for the contractor to prepare his bid is greatly reduced. On a small project, the schedule may be short enough that the time available for contracting will only permit a reimbursable approach.

2. Larger group of bidders. Because it requires relatively little effort to prepare a reimbursable bid, the client can expect that all the desired bidders who have available capacity will submit bids.

3. Can be used in any market condition. When contractors are busy, they generally are willing to quote on reimbursable contracts. When they are slow, they are still glad to quote reimbursable rates, and the client can expect those rates to be substantially reduced. When a contractor's workload drops significantly, cashflow quickly becomes his first priority, and he will often be willing to bid for work at cost or even below cost in order to avoid losing personnel or suffer other unpleasant financial consequences.

4. Allows the scope of work to vary. Since the contractor will be paid for all work done, his scope of work can vary without causing contractual problems. This feature is well suited to the types of small projects whose scope of work simply cannot be well defined at the time the contract is negotiated.

5. Easier to negotiate. Less time and legal assistance is required for negotiation, and standard forms can be used. Many companies negotiate "umbrella" or "general service" agreements under which many small projects can be done.

There are also disadvantages to reimbursable contracts, as seen below.

Disadvantages of reimbursable contracts

The disadvantages of reimbursable contracts stem from the fact that the contractor has every incentive to expend the maximum number of man-hours and incur maximum costs, as every man-hour and purchase on the client's behalf carries with it some extra profit. Unlike the lump-sum contract in which the contractor takes a significant risk, the reimbursable contract puts virtually all the risk on the client. This means that the client has to do an effective job of project control if the desired results are to be achieved. And, in the small-project environment, the manpower, systems, and methods required for effective control may not be available.

The Right Approach

What, then, is the right contracting approach for a small project? For those small projects for which the scope of work can be well defined, the schedule will permit the time required for contracting, and during which market conditions are expected to be favorable, then the lump-sum contract should be considered. For those with a tight schedule and/or a variable scope of work, a reimbursable approach is probably appropriate. The necessary project-control techniques are described in subsequent chapters.

For many small projects, a contracting approach that is a combination of the best features of the lump sum and reimbursable contract may be appropriate, as described below.

Reimbursable Contracts With Fixed Fee

One way to reduce the contractor's incentive to expend the maximum number of man-hours is to fix the fee. In most contracts of this type, the fixed fee is based on the agreed-upon estimate of man-hours required to do the initial scope of work. As changes are issued and the scope of work increases, the fee is adjusted according to a negotiated formula, usually based on the estimate of manhours associated with the change or increase in scope. This type of contract works well in most situations where a reimbursablr approach is required.

For example, suppose it is agreed that the initial scope of work, as initially defined, will take 10,000 man-hours. If the all-in rate per hour is $30, of which the contractor's fee is $3, the fee for the initial scope of work will be $30,000, *regardless of the number of man-hours actually spent.* If the man-hours exceed 10,000, additional man-hours are reimbursed at a rate of $27 per hour; that is, without any fee payment. If additional work is authorized, the man-hours and fee are agreed beforehand.

Unit-Price Contracts

In unit-price contracts, a pricing schedule for performing specific physical tasks is negotiated. For example, if the work involves the welding of pipe the unit-price schedule would contain prices for welding of pipe of various diameters, flange ratings, and materials. As the work is defined and drawings are released for construction, each drawing is assigned a price, based on the quantities shown, the number of tasks required, and the price per task. In this example, an isometric drawing would be priced according to the number of welds required in each category, and the agreed-upon unit-price sheet.

Unit prices can be thought of as lists of many lump-sum prices for very small packages of work. As such, it is like a lump-sum contract in which the scope of work can vary, and is therefore a useful form of contract for the small project. A big advantage is that contract administration and project control focus on physical quantities, not man-hours. One disadvantage of the unit-price contract is the difficulty of comparing bids and selecting the low bidder. Since the pricing schedule comprises many pages of detailed information, the mix (or "balance") of operations which is expected on a specific job will be a key factor in determining the low bidder. The only effective way to determine the low bidder is to make up a sample scope of work and price it according to each bidder's pricing schedule. Each bidder will allocate profit differently to the various operations being priced.

Most unit-price contracts, like most lump-sum contracts, also provide a schedule of hourly rates for use when the work scope is not covered by the unit-price schedule. When work is progressing simultaneously on both unit-price (or lump-sum) work and reimbursable work, the contractor will have to be watched closely to assure that hours covered by the unit price or lump sum are not also being treated as reimbursable.

Negotiated Contracts

When it is determined that there is only one contractor who can perform the required work, the contract—which is then negotiated on a non-competitive basis—is a "negotiated contract." Many small projects are done this way, due to the use of preferred contractors, or those with specialized skills. Despite this, the client need not be in a weak position.

The first rule of negotiated contracts is to avoid jumping to the conclusion that there is only one acceptable bidder. Although it may mean extra work to identify and qualify one or more new bidders, the results may well be worthwhile. Project engineers often go to a single contractor simply because that contractor has done similar work before and is known to be reliable. Although those are good reasons for selecting a contractor, the company's best interests are served when several qualified contractors are available for anything that needs to be done. And, the best way to qualify a contractor is to have him do some work for you, so the assumption that a single-source contract is necessary should be challenged.

In those cases where a negotiated contract is necessary, the same principles should be applied as those used in a competitive situation. The client can estimate what the contract pricing schedule would be if a competitive situation existed by using past contracts and current-cost data as a guide. The client can also ask the contractor to provide "open book" details of his bid, to show how the pricing was arrived at, and why it is fair. Finally, he can use the strong bargaining position he has as the client, possibly an important one, whose continued business is valuable to the contractor.

Incentive Plans

Another means of creating the incentives to perform that are associated with lump-sum contracts is through the use of incentive plans. Although these are generally associated with larger projects having complex contracts, there are many small-project situations where an incentive plan is useful. For example, a critical maintenance or a major turnaround project, in which a great deal of valuable production is lost for every hour the plant remains shut down, can benefit from an incentive plan in which the contractor earns a significant bonus for finishing on or ahead of schedule. Experienced project engineers know that the point of view of the client and contractor are quite different: each makes a profit in a different way, and each therefore views the project from his own profit-making perspective. The cost overrun that could seriously diminish the client's profit could be a boon to the contractor with a reimbursable contract.

The purpose of an incentive plan is to align the objectives of client and contractor by giving the contractor a profit incentive to do what also benefits the client. A good incentive plan should be designed so that it is quite possible for the contractor to earn a bonus. The benefits to the client should be such that he wants to pay the bonus as the cost of the bonus is far less than the financial benefit he derives from improved contractor performance.

Most incentive plans involve schedule performance, although costs and other project parameters can also be included. Incentive plans can be incorporated in any type of contract. There are two general categories of incentive plans.

Unilateral Incentive Plans

Unilateral incentive plans are not negotiated: the client simply makes the contractor aware that he will receive a certain bonus if he meets certain targets. For example, he might offer a bonus for completion on-schedule, with an increased bonus for every day that completion is ahead of schedule (see Figure 3.14) The unilateral incentive plan is relatively simple to administer because there is no negotiation involved: that is, it is a "take it or leave it" offer. As a result, such incentive plans usually provide a bonus for good performance, but no penalties if performance is not up to expectations.

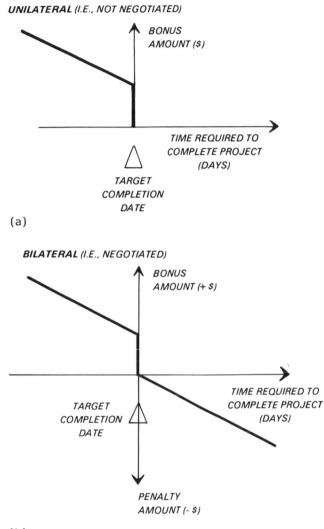

UNILATERAL *(I.E., NOT NEGOTIATED)*

BONUS AMOUNT ($)

TIME REQUIRED TO COMPLETE PROJECT (DAYS)

TARGET COMPLETION DATE

(a)

BILATERAL *(I.E., NEGOTIATED)*

BONUS AMOUNT (+ $)

TIME REQUIRED TO COMPLETE PROJECT (DAYS)

TARGET COMPLETION DATE

PENALTY AMOUNT (- $)

(b)

FIGURE 3.14 Incentive plans. (a) Unilateral (i.e., not negotiated). In this example, the contractor will get a bonus for meeting the target completion date. For every day ahead of the target completion, he receives an additional bonus. There is no penalty for failing to meet the target date. (b) Bilateral (i.e., negotiated)

Bilateral Incentive Plans

These are negotiated incentive plans, in which the client and contractor agree on every aspect of the plan and administer it in close cooperation. This type of plan can involve penalties for poor performance as well benefits for good performance. It can benefit the client, providing greater incentive because of the contractor's involvement in its design and administration.

Cautions on Incentive Plans

Although a well designed incentive plan can produce an outstanding return on the effort and cost involved, incentive plans can also be counterproductive. One problem can be the contractor's preoccupation with the bonus to the point at which it becomes a distraction and too little attention is paid to more important matters. The incentive plan must be carefully designed and administered such that it is difficult but achievable. The benefit is clearly lost if it is too easy, and, if the target is too difficult, a lot of money and effort is wasted in pursuit of an impossible goal.

Steps Involved in Contracting

The project plan should allow ample time and resources to perform the work required for contracting. A properly executed contract plan will contain the following steps:

1. Screening of contractors. This is the identification of those contractors who are both able and willing to bid and perform. This is particularly important when a lump-sum bid is planned. The screening can be done by letter, telex, or even by telephone, using a standard questionnaire.
2. Selection of bidders. Based on the list of potential bidders generated in step 1, the final list of bidders is prepared. Factors in the decision as to which contractors will be asked to bid are:

Performance on past projects
Capability
Current workload
Interest in doing the work

When it comes to bid lists, many project engineers operate on the basis of "the more the merrier." In fact, there is little to be gained by having too many bidders, except a lot of extra work and lost time. While there is no rule as to how many bidders are appropriate in a given situation, the project engineer should be guided by the principle that a Request for Proposal should be given only to those contractors who are best qualified to do the work.

3. Preparation of the Request for Proposal. The request for proposal should be carefully prepared and state, as specifically as possible, what will be required of the contractor both in his bid and, should he get the contract, in his performance of the work. The more the contractor knows of what is expected, when bidding, the less excuse he has later for not performing. The Request should in all cases contain:

A description of the scope of work (general for reimbursable contracts, specific for lump sum)
The reporting requirements for project control (discussed in subsequent chapters)
The specific format for the bid pricing (to assure that bids can be compared on an consistent basis)

4. Bid analysis. This involves the reviewing of the bids to select the preferred contractor. Bid review is enhanced considerably when a check estimate has been prepared. Each bid can then be compared in every respect to the check estimate, and errors, omissions, or things which seem out of line can be identified and, if necessary, corrected. The bid review really consists of two separate analyses:

"Hard-money analysis": This is the quantitative analysis of pricing. compliance with contract items, and other commercial aspects.
"Soft-money analysis": This is the qualitative analysis of contractor capabilities, based on past performance, quoted performance, and specific evaluations by the client. Using a rating sheet (much like a report card: see Figure 3.15) the relative technical and management capabilities can be ranked.

In some cases, the choice turns out to be difficult, such as a contractor whose performance is predicted to be above expectation, but whose price is not the lowest. Anyone who has ever tried to save money by buying the cheapest product knows that the low bidder is not necessarily the best choice, but this does not make it easy to explain to management why the low bidder was not chosen. One way to quantify the judgement process in such cases is to use the network model. For example, the network durations can be changed to reflect the anticipated schedule performance of each contractor, and the net effect on schedule of selecting each contractor can be evaluated. Then, the economic benefits of the improved schedule performance can be compared with the increased contract costs. Or the economic model used to justify the project can be run to simulate the results of using different contractors with varying design capabilities, by varying operating costs and service factors.

USE FOR CONTRACTOR SELECTION AND/OR FOR PERFORMANCE DATA

RATING SCALE: 5 = EXCEEDS ALL EXPECTATIONS
              4 = ABOVE AVERAGE
              3 = AVERAGE
              2 = BELOW AVERAGE
              1 = UNACCEPTABLE

		CONTRACTOR		
CAPABILTY	WEIGHT	A	B	C
PRODUCTIVITY	0.20	4	3	2
DESIGN	0.10	4	2	3
PERSONNEL	0.25	4	4	3
FACILITIES	0.10	3	3	3
RESPONSIVENESS	0.15	2	4	3
PAST PERFORM.	0.20	3	4	3
NET RATING	1.0	3.4	3.5	2.8

FIGURE 3.15   Contractor rating sheet.

    5. Contract award. This final step involves the negotiations with
the selected contractor. This is the last chance for the client to negotiate from a position of strength; once the contract is signed the contractor, as discussed earlier, has the stronger position. Many clients
overlook the importance, at this point, of defining the reporting requirements that will assure project control: such things as progress
measurement, manpower, reporting cycles, etc. *The essence of effective contractor control is the contractual obligation to provide information which can be used by the client for unbiased performance measurement.* Such information is seldom offered voluntarily.

PLANNING FOR CONTROL OF TECHNICAL QUALITY

Project control, as pointed out in Chapter 2, is really a process of optimization of the cost, schedule, and quality to achieve the maximum
profitability. Although most of this book deals with the control of cost
and schedule, quality control is an equally important aspect which affects them both (see Chapter 9 for further discussion).

Technical quality is a subject in which many words have been used interchangeably to the point where their meaning has been lost. As before, the definitions offered here are those which have found general acceptance, and which will be used throughout this book.

What is "quality"? From a project engineer's standpoint, the relevant terms are defined as follows:

Quality: fitness for the intended purpose. Using this definition, it can be seen that the maximum quality for an engineered item is not necessarily the use of the most expensive materials, or the latest technology, or hand-fabrication. The only attributes which add to quality are those which make it more fit for its intended purpose, which is usually to enhance profitability. This simple definition, when applied to proposed design changes, can help separate those that are necessary from those that are not.

Quality assurance: the method by which the work will be controlled to assure satisfactory quality. On most projects, quality assurance requires a definition of quality in terms of specific standards (such as building codes, API specs, etc.) which will be enforced. A recent trend in defining quality in specific design terms is to consider it in terms of life-cycle costs. Since some performance standard is expected of every part of the facility over its entire operating life, it is reasonable to look at design vs. cost evaluations not only in terms of investment cost, but also in terms of the impact that a design change could have on operating costs such as power consumption and maintenance. The first step in quality assurance is, therefore, to define design quality in specific performance terms, as well as in terms of design criteria.

Quality control: the actions which assure acceptable quality. Typical quality-control activities include inspection and testing of manufactured parts and assemblies, welding inspection, certification (when appropriate), and design reviews. Design reviews are a recent trend which has been shown to aid significantly in reducing late changes and assuring fitness for the intended purpose. Design reviews require that the ultimate users of the facilities—usually operations and maintenance staff—participate in periodic reviews of the design and construction, be made aware of current design problems and decisions, and make any input to the design process that they feel appropriate. If certifying bodies are involved, their reviews must also be scheduled.

If effective quality control is to take place, the necessary activities must be shown on the network plan and schedule. This is another project-management function which is often overlooked in the planning phase, causing problems such as late changes and rework. Typical quality-control activities which should be included in the plan are:

Design reviews
Reviews by certifying authorities (if applicable)
Visits to fabricators and/or witnessing of tests

CHAPTER SUMMARY

This chapter described the techniques available for network planning and scheduling. Important considerations which affect the plan, such as contracting strategy and quality assurance, were also described.

The network plan and barchart schedule which have been developed for small projects, using the methods described so far, is the foundation of all that follows. Scheduling the use of resources, cost estimating, requesting project approval, obtaining commitments of resources, and implementing project control will all be based on this project plan. Although our plan and schedule may change as the project progresses, this original plan will always be our basis for comparison. We will always measure where we are by our distance from it. This original plan is our "static model" of the project.

4
Resource Planning for Small Projects

WHAT IS RESOURCE PLANNING?

Resource planning is based on the inarguable premise that work cannot be accomplished without four essential resources necessary to accomplish the given scope of work: materials, people, equipment, and time. The network plan and schedule developed in Chapter 3 established the availability of the resource time. If the project plan and schedule are to be achieved, it is now necessary to assure that the required material, labor, and equipment will be available when needed and in the required quantities. The process by which this is accomplished is resource planning.

Although the importance of resource planning is self-evident, many projects, small and large, suffer avoidable delays from inadequate resource planning and control. This is especially true for small projects. So one of the most useful applications of our small-project network plan and schedule is that it makes it possible to do a credible job of resource planning. Resource planning is the process of identifying the quantities of resources required to accomplish the work and scheduling these resources over the time of the project, i.e., determining "what" is required and "when." Resource planning consists of the following steps, each described later in the chapter.

1. Define the resources required to accomplish each activity on the network plan. We will refer to this process as *resourcing the network.*
2. Define the total amount of resources required for the project, in each time period, according to the plan and schedule. We will refer to this process as *resource aggregation.*

3. Define the resources which can be made available during each
    time period, i.e., *resource availability.*
4. Compare resource requirements with resource availability to
    identify shortfalls, and reschedule the network plan to assure
    that resource requirements will not exceed availability. This
    is *resource levelling.*
5. Prepare a schedule for the provision of project resources. This
    resource schedule, for manpower resources, is usually presented
    as a *manpower histogram.*

Resource Planning Is Essential for Small Projects

Although resource planning may seem an elaborate process to perform
on a small project, it is, in fact, quite a vital task. Resource planning
is especially important for small projects, for the following reasons:

Small projects are profoundly affected by resource shortfalls due
to the lack of alternative ways to make progress. On a large project, if a certain material is not delivered on schedule, thereby
delaying work in one area, there are usually many other work
areas for which material is available and in which physical progress
can be made. Similarly, the lack of a special piece of construction
equipment, or of certain labor crafts, can be compensated for by
working on other things. Since the large project has many noncritical activities, these adjustments to the plan can often be
made with no net effect on the completion date. These alternative ways to make progress, when resource shortfalls are
experienced, often do not exist for small projects. A resource
plan helps to avoid shortfalls, but, when shortfalls are experienced, the resource plan makes it easier to identify alternative
courses of action and changes to the plan to minimize the damage.
Small projects, which are often executed without any formal planning,
can often suffer delays due to resources not being available when
needed simply because they have not been identified. A resource
plan assures that all necessary resources are identified well in advance, and commitments obtained to provide them on schedule.
Small projects are often more likely to experience resource shortfalls
than larger projects. In an operating plant, where project work
may not be a top priority, resources such as plant labor, equipment, or warehouse supplies may be diverted to higher priority
uses just when they are needed on the project.
Project engineers responsible for small projects often find it difficult
to obtain the necessary commitments from other parts of the organization to provice resources. The resource plan represents a clear
commitment which is agreed upon before the project begins, and
which can be compared with the actual performance. Then, when

promised resources are denied, the project engineer can show the effect that the resource shortfalls have on the schedule: this can help prevent the shortfall from occurring in the first place.

Resource planning provides a good basis for estimating techniques which are appropriate for small projects (see Chapter 5).

DETAILS OF RESOURCE PLANNING

Step 1: Resourcing the Network

Each activity in the plan is reviewed to determine the resources required in each category to accomplish the specified scope of work in the time provided by the activity duration. The easiest way to accomplish this is by filling out a chart as shown in Figure 4.1. It should be noted that the duration is, of course, itself a direct function of the amount of resources applied. When planning networks have been used for some time, a database of actual project experience with manpower and durations can be set up and used as a guide to determine the time and resources required to perform typical tasks.

For small projects, it is generally assumed that the number of resources in each category is constant over the duration of the activity. Where this is not the case, a resource profile for each activity can be established. This type of analysis poses no special problem, but it requires a more sophisticated computer system than is apt to be available for the small project. Proceeding activity-by-activity, the network-resourcing worksheet is filled out. It might seem that, if we were to stop the resource analysis at this point, we would have a reasonable resource plan for the project. However, the next step often reveals that, when total project resources are considered, there are often sharp peaks in the resource requirements that would certainly not be met in actual practice and which can easily be removed by rescheduling.

Step 2: Resource Aggregation

Given the number of resources in each category required to complete each activity, and, given the scheduled time period for each activity, it is a simple matter to calculate the total resource requirement for all activities occurring during each time period. In this manner a histogram of total project manpower can be obtained, as well as one for total manpower in each discipline (see Figure 4.2 and 4.3). Resource aggregation can quickly become a chore when done manually, but there are a number of simple computer programs available—even for personal computers—which can be used. The result of resource aggregation, showing total manpower requirements per time period, is a very useful tool for a number of purposes.

		ENGINEERS			FIELD FORCES					EQUIPMENT	
NO.	ACTIVITY DESCRIPTION	DURATION	DESIGN	PROJ.	LABORER	WELDER	IRONWORKER	ELECTRICIAN	BACKHOE	WELD.MACH	CRANE
1230	PREPARE FLOWSHEETS	6	2	1							
2230	PREPARE P & ID'S	4	3	1							
3230	RELEASE EXCH. FND. DWGS.	1		1							
4230	RELEASE INSTRUMENT DWGS.	1		1							
5230	RELEASE PIPE SUPPORT DWGS.	1		1							
6230	ORDER INSTRUMENTS & C.V.'S	1		1							
7230	PREPARE ISOMETRICS	13	2	1							
8230	ORDER EXCHANGERS	1		1							
9230	DELIVER INSTRUM. & C.V.'S	4		1							
10230	ORDER PIPING	1		1							
11230	FABRICATE & DELIVER EXCH.	21		1							
1308	EXCAVATE EXCHANGER FNDN.	2		1	6				1		
2308	POUR EXCHANGER FNDN.	7		1	4						
1311	SET EXCHANGERS	1		1	3	1					1
1313	FABRICATE PIPING	10		1	10	10				4	
2313	ERECT PIPING	8		1	8	8				4	
3313	HOOKUP & TEST PIPING	2		1	6	4				2	
4313	STARTUP	2		1	4	1		4		1	
1318	FABRICATE PIPE SUPPORTS	5		1	1	1	1			1	
2318	ERECT PIPE SUPPORTS	3		1	1	1	3			1	
1322	INSTALL INSTRUM. & C.V.'S	3		1	2			4			1
2322	TEST INSTRUM. & C.V.'S	1		1	1			2			

FIGURE 4.1 Summary of labor resource requirements.

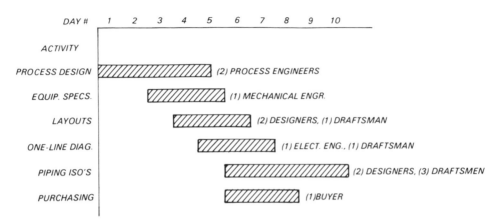

FIGURE 4.2 Resource aggregation. In this example, the bar chart shows the design of a small process-plant project.

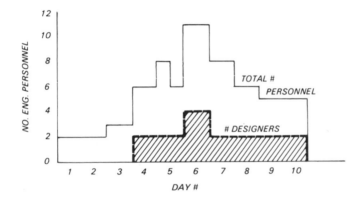

FIGURE 4.3 Resource histograms derived from Figure 4.2.

1. The total manpower histogram is essential in determining the scope of the job in terms of manpower and therefore in scheduling the project itself within the population of other small projects. We can see that the project shown in Figure 4.2, which requires two process engineers to start, can begin as soon as preceeding projects release them.

2. The total manpower histogram can be converted to a cumulative manpower curve, or S curve, as shown in Figure 4.4. The S curve for manpower is useful in project control as a means of comparing the actual expenditure of manpower against the original plan.

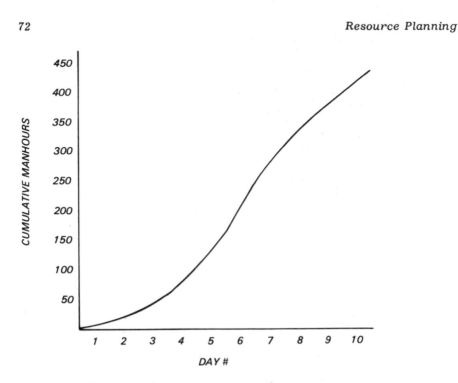

FIGURE 4.4 Cumulative manpower vs. time.

3. The cumulative manpower curve can also be used to derive the S curve for physical progress. This curve, showing cumulative progress as a function of time, is probably the most important project-control device, and is an important part of the plan. If we make the reasonable assumption that productivity is constant over the life of the small project, the progress curve will have the same shape as the manpower curve, and is derived simply by assuming that percentage of planned man-hours expended at a certain point in time is equal to the percentage of planned progress. (Note that we are dealing here only with planned, not with actual, man-hours and progress. This subject is explored in much greater depth in Section III.)

4. The cumulative manpower also provides a good basis for calculating expenditures, since most of the cost of a small project is likely to be labor-related. Expenditure forecasting is a particularly time-consuming task which is important but contributes little to project control, and any short-cut approaches are usually very useful.

5. The cumulative manpower and progress curves are also useful for small projects because they often do not resemble the traditional "S" shape which is generally seen on large projects. The S shape results

from the fact that most large projects begin with activities, such as civil work, which must precede the rest of the job and yet can only utilize a certain number of people. Once the sitework is done, the underground piping and ducts installed, and the foundations complete, then many other activities can commence. Similar phenomena also exist toward the end of the large project.

Since many small projects, such as turnarounds, maintenance, and other in-plant projects, start with a full workforce and work in many areas at once during the project's short duration, the shape of the cumulative manpower and progress curves is impossible to predict without some sort of resource aggregation procedure.

Step 3: Defining Resource Availability

Manpower Availability

The availability of manpower resources can be limited by a number of factors including:

Manpower availability: Small projects which must draw manpower resources from a labor pool which is shared with other projects (e.g., drawing office, plant mechanical forces, or fixed company staff), and which must rely on manpower to be released from preceeding projects, will be likely to experience clear limitations in manpower availability.

Manpower density: Many types of construction work are limited by the number of workers per square foot that can work efficiently. When manpower densities exceed the guidelines, productivity is likely to suffer due to interference, supervision, and logistical problems. Safety can also be a reason to set limits on manpower density.

Physical constraints: Some operations, such as those which occur above grade, can only absorb a certain number of workers due to physical limitations. For example, the erection of the frames for a heavy steel structure can only utilize a certain number of steelworkers. Or, if work must be performed on scaffolding, the extent of scaffolding installed can be a factor in limiting manpower.

Limited availability of special skills: Some engineering or construction projects require skills which are in short supply, either inhouse or from contractors. This problem is particularly acute for projects in remote locations, where the local population cannot supply the needed skills, and the number which can be transported in is limited.

Logistics due to project location: Some projects, best exemplified by offshore oil platforms, have severe limitations placed on the

availability of manpower simply by the logistics involved. Those
who have worked offshore know that the number one factor in determining manpower availability is the number of beds available on
the platform. Other logistical factors such as supply boats and
helicopters play a big part in determining the manpower available.

Material and Other Resource Availability

Limitations on availability also exist for material and other resources.
Material resources on small projects can be limited in availability due to
delivery schedules, warehouse stocks, or project priority. Small projects often suffer from material shortages caused by preceeding projects taking materials intended for the project at hand. Remote projects, of course, have material availability problems stemming from the
logistics involved in receiving, storing, and transporting materials to
remote sites. In some cases, such as offshore projects, storage on-site
is limited as well.

Construction equipment availability can be limited by the requirement
that in-plant equipment be used, and the necessity of sharing the pool
of available equipment with other projects. Often the failure to properly define project priorities can lead to heated debates about where
a particular piece of equipment ought to be used next. Equipment
availability can also be limited by the physical constraints of the site,
as well as by the need to avoid congestion and interference with the
operating plant.

After careful consideration of the above factors, and discussions with
those who wil be responsible for providing the resources, the project
engineer is able to draw a histogram for resource availability in each
category as shown in Figure 4.5. By overlaying the resource-avail-
ability histogram on the resource-requirement histogram, the resource
shortfalls in each category can be seen.

For the project to be completed according to plan, it is now necessary
to adjust the schedule to bring resource availability into line with requirements.

Step 4: Resource Levelling

The process of resource levelling, usually done by computer, proceeds
as follows:

> The *early-start schedule,* in which each activity is assumed to start
> on it's early-start date, is *used for the initial resource aggregation,*
> By comparing requirements with availability, resource shortfalls
> are identified.
> The first attempt to eliminate resource shortfalls is made by delaying
> the start of various activities. This can be thought of as allowing

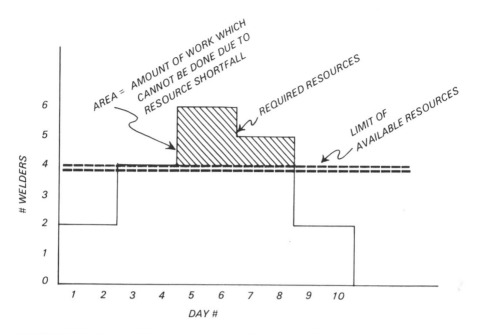

FIGURE 4.5 Comparing resource requirements with resource availability.

noncritical activities to move within the time envelope of their duration plus float (see Figure 4.6).

If resource shortfalls still exist, it is necessary to allow either the schedule for the project to slip or to increase the available resource levels (see Figure 4.7). If the project-completion date

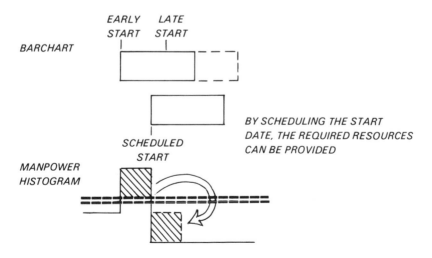

FIGURE 4.6 Using float to eliminate resource shortfalls.

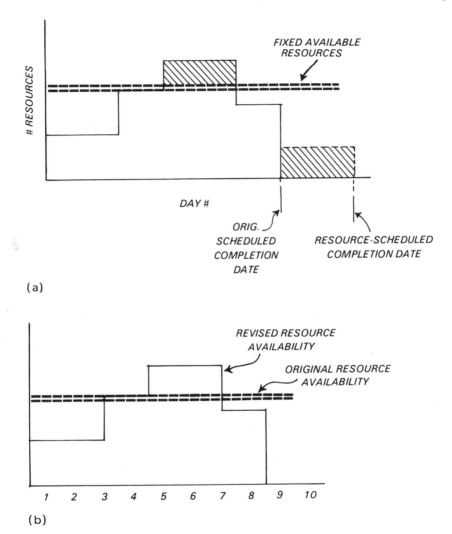

(a)

(b)

FIGURE 4.7 (a) Resource- and (b) time-limited scheduling. Resource-limited scheduling allows the schedule to slip to match availability while time-limited scheduling allows resource availability to match requirements.

FIGURE 4.8 Schedule of resource requirements. (a) Project engineers, (b) Designers, (c) Laborers, (d) Welders, (e) Mobile cranes, (f) Welding machines.

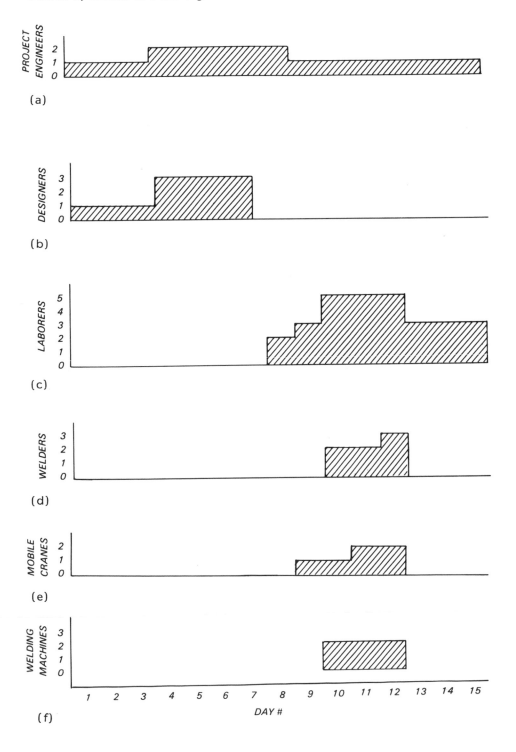

cannot be allowed to slip, we have a case of *time-limited* resource
levelling, in which we simply identify the time period and amounts
that various resource availabilities must be increased. If the resource availability cannot be increased, we have a case of *resource-
limited* levelling in which we must allow the schedule to slip in order
to accomodate the available resources.

From the above it can be seen that resource levelling, like scheduling, is very much an iterative process in which the goal is to find the
optimum combination of the project parameters of time and resources.

Step 5: Schedule Resource Requirements

With the resource levelling complete and the resourced schedule final-
ized, it is appropriate to prepare a schedule of requirements for those
who will be responsible for providing the necessary resources. These
resource schedules can take the form of a histogram or even a simple
list showing what is required and when. The resource schedule should
then be used to obtain agreement and a commitment to provide the resources as scheduled. Examples are shown in Figure 4.8.

ALLOCATING RESOURCES TO MULTIPLE
PROJECTS

One of the biggest problems of small projects is the management of
many projects at once. Perhaps the greatest difficulty involved in
managing multiple projects is that of deciding how to allocate resources
among them. For example, the manager of the project-engineering department may have a "pool" of 10 project engineers, each of whom is
handling 10 projects. A new project comes into the department, and
he must decide to whom this project should be assigned. Or, the
drawing-office manager may have just so many designers and draftsmen: how should they be allocated to the various projects? The
project engineer has a similar problem where his own projects are
concerned. These resource pools usually also exist for construction
labor and equipment, as well as for materials.
The solution to this problem lies in the concept of network heirarch-
ies, as discussed in Chapter 3, combined with the principles of re-

source aggregation. After resource planning on an individual basis at Level 2, the project can be represented, using hammocks, as a simple Level 1 network, whose resources are summarized. The several projects under consideration can then be handled as if they were diverse parts of a single project, which, in fact, they are. The same principles of resource allocation can then be applied to the family of projects at Level 1.

In the example shown in Figure 4.9, we have 10 engineering projects, each represented by one or more activities on the master network. This master network represents a "project" which consists of all the current and planned small projects we are managing. For example, it might represent the current plan for all the projects to be done this year. Although many projects are independent, there will be some cases in which one project must be complete before the next one can start, and the network reflects those constraints. What, then, will determine when each project gets done? There are usually three factors:

The project's priority
The constraints shown in the network
Resource availability

While the priorities and constraints may be out of the project engineer's control, he can control how the resources are deployed. If he performs a normal resource analysis on this multi-project network, resource aggregation will indicate which projects are likely to be delayed because of shortfalls, and resource levelling will indicate how the projects can be scheduled to match the available resources in the pool. When new projects are added and completed, and when priorities or the availability of resources are changed, the analysis can simply be updated. Such an analysis can be useful in justifying the need for outside services, such as subcontracted engineering work. Many project-management software packages offer this multiproject capability.

This technique is particularly useful for planning and decision making at the management level. It can be used to address problems caused by several projects competing for limited resources, problems that arise when the available resources for a project have to be increased, and also to help set priorities.

NETWORK — EACH PROJECT REPRESENTED BY ONE OR MORE ACTIVITIES

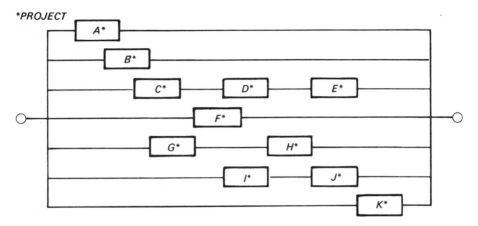

FROM THE NETWORK, A BARCHART SCHEDULE IS
PREPARED ... AND RESOURCES ALLOCATED TO EACH PROJECT

(a)

RESOURCE ANALYSIS (EXAMPLE: ENGINEERING PROJECTS)

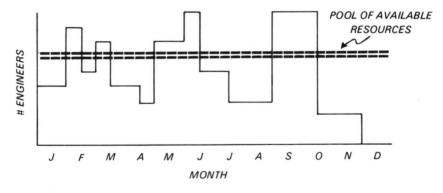

RESCHEDULING ... TO REFLECT:
— PRIORITIES
— SEQUENCE
— RESOURCE LIMITATIONS

(b)

FIGURE 4.9 Allocating a fixed pool of resources to multiple projects.
(a) The network shows each project represented by one or more activities. A, B, C,...,K represent these projects. From the network, a
bar-chart schedule is prepared and resources allocated to each project.
(b) Resource analysis (in this example: engineering projects). Rescheduling reflects priorities, sequence, and resource limitations.

CHAPTER SUMMARY

In this chapter we discussed the importance of resource planning to the small project, and showed how it should be done. We also accomplished the first step in integration, by loading labor and material resources onto our network plan. The original schedule may have changed as a result, but we can now be confident that we know what resources have to be provided so that the plan can be executed according to schedule.

5
Cost Estimating for Small Projects

IMPORTANCE OF THE COST ESTIMATE

The cost estimate for a project serves two important functions:

> It is the basis on which the *economics* of the project are calculated,
> and therefore an essential element in the decision to go ahead.
> It serves as the basis for *cost control.*

In spite of the importance of the cost estimate, most small projects suffer from a lack of accurate, consistent, well-organized estimating. As mentioned in Chapter 1, this is often due to the inherent difficulty in estimating small projects involving revamp work, as well as to the lack of trained estimators and formal methods and data. Unfortunately, small projects have a tendency to overrun, due to variations in the scope of work, unforeseen difficulties, etc., and, when they do, the overrun percentage is apt to be great because the original estimate is small. To put it another way, management often fails to realize that it doesn't take much to cause an overrun of greater than 10% on a small project.

It is, therefore, evident that a better way to estimate small projects is needed: one which will provide greater consistency and simplicity; provides a framework for capturing actual cost data for use in subsequent estimates; addresses the difficulties of estimating labor on revamps; and provides a good basis for cost control. This chapter describes the concepts and methodology necessary to accomplish those goals.

THE INTEGRATED APPROACH TO ESTIMATING

Chapters 3 and 4 describe the methods by which we create a plan and schedule for the project that defines:

The *activities* which must be performed in order to complete the project
The *sequence* in which the activities will be performed
The *time period* in which each activity will be performed
The *resources* which will be applied to each activity

After the planning, scheduling, and resource levelling have been completed, all the project parameters are in balance, that is, the resources that will be applied are adequate to accomplish the defined scope of work of each activity in the scheduled time. It now remains to estimate what it will cost to provide those resources. It is therefore appropriate to use the resource plan as the basis for estimating labor and construction equipment costs.

The basis of the estimate in the integrated approach to cost estimating derives from the resource-loaded project plan. For each activity, the cost is estimated:

To do the *work* required
In the allowed *time*
With the allocated *resources*

This technique is particularly appropriate for small projects as the labor-related costs represent a major and often ill-defined part of the project cost. The estimating process is also greatly simplified, and is well suited to situations in which a lack of relevant historical data is available and judgement must therefore be used.

The integrated approach wil be described in detail in this chapter. However, the basic principles of cost estimating also apply to small projects, and these are discussed below.

PRINCIPLES OF COST ESTIMATING FOR SMALL PROJECTS

What Is a Cost Estimate?

Before exploring further the concepts and techniques of cost estimating, it is appropriate to define exactly what we mean by a cost estimate in the context of a project. A cost estimate is:

1. "A calculation of the *approximate* cost." This definition, from Webster's dictionary, reminds us that, contrary to what is popularly believed by those who approve budgets, and estimate is, by definition, an approximation. When functioning as estimators we need to remember this and make sure those who work with our estimates remember this too. There are ways to measure the accuracy and uncertainty in an estimate, and these will be discussed later in the chapter, as well as in Chapter 10.

2. A *forecast* of the final cost for the project, under an assumed set of conditions. This part of the definition reminds us that an estimate is based on assumptions. Even if we have some "firm data" on design or cost, we are assuming that firm data will remain firm for the life of the project. For our estimate to be exactly correct, every assumption, whether explicit or implicit, must prove to be correct, and this very seldom happens. So, we should look for an estimate to be approximately correct, not exactly correct. All the basic assumptions should be reasonably close to correct, and those areas which were underestimated should be offset by those which were overestimated.

This definition also reminds us that an estimate is a *prediction*, or forecast, of what will happen over the project's life. The extent to which information is undefined at the time of the estimate is predicted by the estimate according to the assumptions made and the methods used. Estimate accuracy increases as the amount of definition increases, and the amount of prediction decreases (see Figure 5.1).

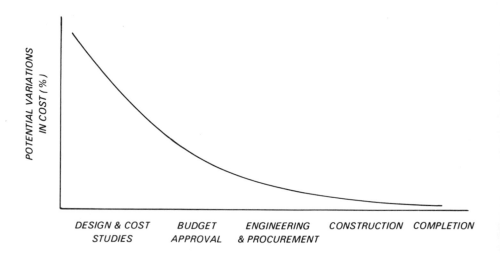

FIGURE 5.1 Estimate accuracy improves with project definition.

3. A *reflection of the plan* for the project's design and construction. The cost estimate for the project can be thought of as a "budget" in the literal sense of the word. In financial planning, the word budget implies "plan"; that is, the budget is the plan for how money will be spent. Our project-cost estimate is also such a plan, and, as such, must reflect to the greatest extent possible how the work will be done.

Characteristics of a Cost Estimate

Unbiased

Unless stated otherwise, the estimate should have an equal probability of overrun or underrun. This requirement is especially true for small projects. The reason for this is simple: if a company is funding 100 projects per year, and, if each is estimated at the 50/50 probability point, the odds are that the money set aside for all the projects will be about right. If, however, all the projects have conservative estimates, with extra contingency, the odds are that too much money will end up being set aside for the projects. As a result, the unspent project funds will not have been available for other investments or expenditures. Anyone who has completed an internal project at a large cost underrun and expected thunderous acclaim has probably found that the reaction to having the funds returned was not so good, since it was correctly perceived that those funds could have been put to good use elsewhere.

In the case of estimates for bidding, or for particularly cost-sensitive projects, the 50/50 estimate may not be the most desirable. In the case of a bid, the probability of getting the work must be weighed against the probability of overrunning the estimate. In the case of a cost-sensitive project, the project may be necessary but risky so it is therefore desirable to set aside sufficient funds so that there is a low probability of overrunning. In either case, however, it is important to assure that any bias away from the 50/50 probability point is clearly documented and understood.

Appropriate to the Project Definition

Many companies make the costly mistake of creating an estimate at a level of detail which is far greater than the level of reliable design information available. This practice not only produces an estimate of greater cost and only equivalent accuracy, but it also fosters the dangerous illusion that the estimate is more accurate than it really is. The fact is, an estimate can be no better than the information on which it is based. If a project is in the early design stage, the preliminary nature of the design basis makes any estimate prepared at that stage a preliminary estimate, no matter how much spurious detail may be included. Therefore, the method used, the time and cost, and the basis of an estimate should all be appropriate to the project's status and to the purpose for which the estimate is intended. This is illustrated as follows:

Early-planning estimate

This kind of estimate might be used to make an initial evaluation of a project, or to rank potential projects and select those for further work. Relative accuracy is more important here than the absolute value of the estimate, so quick, curve-type estimates are appropriate. Other techniques could include estimates made by adjustments to data from previous similar projects, as well as those based on "rules of thumb."

Budget estimate

This kind of estimate is used to set the budget for the work, obtain management approval, and acts as a basis for cost control. If detailed design information is available, the estimate should be done in the same level of detail. If not, the estimate should reflect the information available which is considered firm.

Establishing the Estimate Basis

One of the most important concepts in cost estimating, especially where small projects are concerned, is establishing the basis of the estimate. A cost estimate can be thought of as resting on a "tripod" (see Figure 5.2). Each "leg" of the tripod is described as follows:

The Design Basis: What Is to Be Built

The design basis usually takes the form of drawings, sketches, and specifications. The most significant things about the design basis are the level of definition and the potential for change. Many small projects, especially those involving modifications to existing facilities, are defined in great detail by the design group and appear to be on a firm basis. However, once the project progresses further and revamp problems become apparent, major scope changes are likely to occur. For example, the original design often assumes that facilities can be installed in the most straightforward way. Sometimes the facts turn out otherwise. Therefore, the potential for change is as important as the level of detail of design definition.

Figure 5.2 also illustrates that the amount of definition of the basis determines the estimate accuracy: the more that is defined, the less the estimate must predict, and the more "solid" the basis.

The estimate will be correct to the extent that it correctly reflects the design basis and predicts the extent of changes.

The Planning Basis: How It Is to Be Built

The planning basis usually consists of the network plan, barchart schedule, and resource plan as defined in Chapters 3 and 4. The

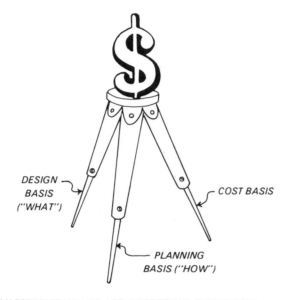

ALL VARIATIONS IN PROJECT COSTS CAN BE EXPLAINED
BY VARIATIONS IN THE DESIGN, PLANNING & COST BASIS

DESIGN BASIS ("WHAT")

COST BASIS

PLANNING BASIS ("HOW")

AN ESTIMATE WILL BE ACCURATE TO THE EXTENT THAT
IT CORRECTLY PREDICTS THE UNDEFINED PART OF THE BASIS

(a)

FIGURE 5.2 The project cost stands on a "tripod" base. All variations in project cost can be explained by variations in the design, planning, and cost basis. (b, left) The early planning estimate must predict most of the basis; (b, right) The detailed estimate has most of its basis defined.

relationship between how the project is managed and its cost has now been discussed sufficiently that it can be seen that the estimate should reflect the contract plan as well as any costs associated with maintaining or accelerating the schedule.

The estimate will be correct to the extent that it correctly reflects the planning basis and predicts how the project will actually be built.

The Cost Basis: What Pricing Levels Will Be Experienced

The cost basis is the pricing used in the estimate. It consists of the data used for estimating, the assumptions made as to escalation, productivity, and other estimating parameters, and the judgements made

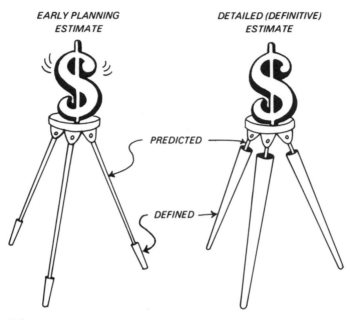

(b)

FIGURE 5.2 (Continued)

by the estimator. Since most estimates are done at some base date and then escalated, the basis should also define the base cost level.

The estimate will be correct to the extent that it correctly reflects cost data and predicts what pricing and performance will actually be experienced.

The estimate basis should be documented carefully and, if possible, be approved before the estimate is prepared. Assumptions, in particular, should be well documented, as well as calculations based on judgement. Documentation is important to enable the project engineer to reconcile the cost estimates and forecasts (which will subsequently be prepared) with the original budget, by studying variations to the estimate basis (see "Reconciling Forecasts and Estimates" and "Estimate Preparation," later in this chapter).

The Estimate Is the Basis for Cost Control

Like the plan and schedule for the project, the estimate's most important function after project approval is to permit cost control. In order to be effective as a control device, the format of the estimate must be closely matched to the way the project will be run. For example, it

should be easy to group costs into the major contracts and subcontracts, and to use the cost codes for aggregation and reporting of cost
data from contractors and other sources. The integrated approach
serves this purpose well. In cases in which the estimate used for project budgeting and approval is not broken down into sufficient detail
for control, many companies find it is worth their while to create a
control estimate with the necessary detail.

Reconciling Forecasts and Estimates

A very useful tool in cost control which is frequently overlooked is the
use of reconcilations. An estimate reconciliation is simply a comparison
of a past estimate or forecast to a current one, and an explanation of
the differences. Through this process, trends can be brought to light
which might otherwise have been overlooked. In the small-project
environment, reconciliations can be a good shortcut method for keeping
track of things and keeping the right people informed. Management,
which must always be conscious of the budget, always appreciate good
reconciliations. Unfortunately, good reconciliations are seldom seen,
because the principles of reconciliation are not widely understood.
Useful reconciliations compare:

Current-cost estimates against the previous estimate (this indicates
major design changes)
Current-cost forecasts against the previous forecast (this indicates
recent trends)
Current-cost forecasts against the current-control estimate (this
indicates forecast trends)
Current-cost forecasts against the budget (this indicates changes
and developments since project approval)

Of these, the two most useful compare the current forecasts or estimates against the previous forecase or estimate (i.e., how have things
changed since we last looked at the cost?) and current forecasts
against the budget.

A useful form of a reconciliation is shown in the following example:

Cost (1,000s)	
(Previous estimate)	(100)
Design variations	+14
Added piping	+6
Revised electrical	+5
Alloy materials	+4
Net all other design variations	(1)

Cost (1,000s)	
Planning variations	(5)
Subcontracted drafting	(3)
Overtime hours	+2
Shared field facilities	(4)
Estimating variations	
Equipment pricing	+2
Market conditions (piping)	(1)
Steel quantities	+4
Contingency reduction	(9)
Net all other	+3
(Current cost forecast)	(108)

It can be seen that this reconciliation format shows the variations to the project cost, broken down in terms of the three legs of the estimate basis: the design, the planning, and the cost basis. All variations to the estimate fall into one or more of these categories. A reconciliation often shows that, although the total cost has not changed significantly, the variations in the basis are quite significant and could be a sign of greater variations to come. In addition, the nature of reconciliations is such that they reveal problem areas which might not otherwise show up.

In preparing a reconciliation, the calculation of the cost impact due to each variation is done on the same basis as the original estimate and assumes that nothing else has changed. Thus, each line should show how the original estimate would have been different if the single variation at hand had occurred.

DEFINITION OF ESTIMATING TERMS

As in planning and scheduling, the proper definition of terms is most important in cost estimating. Unfortunately, the terms used in cost estimating are used so often in other contexts, that they are often ambiguous without this being recognized. It is therefore essential that we begin our discussion of estimating techniques by defining the appropriate terms. As before, though the definitions offered here may not be the only correct ones, they are definitions which are widely used and which will be used in this book.

Project-cost estimate: This is the estimated cost to complete the scope of work defined in the project. This definition can be difficult when the question becomes, "Where does the project begin and end?" The cost estimate should clearly specify the scope of work by stating the design basis and by specifying inclusions and exclusions so that it is clearly understood what is and what is not covered by the estimate.

Capital costs: Many, if not most, projects are "capital projects," that is, they result in a facility which is treated as capital, or as a "fixed" asset in the financial sense. The definition of capital or fixed assets in a company often provides an indication of what should be included and excluded in the capital cost estimate.

Expense costs: These costs are those which can be considered to be necessary for the operation of the existing facilities. The confusion which often exists over capital and expense costs comes from the fact that most projects include both types of costs, as well as from the fact that the tax implications of these costs adds a level of complexity beyond the experience of most project engineers. For example, the costs to build a new facility are clearly capital costs, but the costs to start it up are often considered expenses. There is often a good deal of confusion on a given project as to how these costs should be classified, and the project engineer needs to have these questions resolved if effective costs control is to take place.

Direct material costs: This is the cost of the material which is required for the fixed project facilities, is not consumed during the project, and is not reused elsewhere. Piping, vessels, buildings, and pumps are examples of direct materials. Catalyst, replacement parts, warehouse spares, scaffolding, concrete forms, and welding rod are examples of indirect materials. It is evident from the above that direct materials are in general, capital costs, and indirect materials tend to be expense cost.

These costs are generally divided into two categories:

1. *Equipment* refers to those items which produce some process change. For example, pumps, heat exchangers, pressure vessels, and compressors all have, as their purpose, the changing of pressures, temperatures, and other process variables. In a processing facility, such items as generators and transformers would not be considered equipment because they are part of a utility system. However, in a company which builds power systems, the transformers and generators would be considered equipment. Once again, the important thing is to define the terms for company use and then stick to a consistent definition.

2. *Bulk materials* (sometimes referred to as materials) covers those items which are necessary for the equipment to function. Examples of bulk materials are piping, electric cable, foundations, and structural steel.

Direct labor: Another area of great ambiguity is that of direct and indirect labor. Direct labor is the labor which produces measurable physical progress. This definition intentionally rests on the method used to measure progress. One frequent problem with most current practice is that people compare actual man-hours spent with measured progress and draw conclusions: if the man-hours and progress measurement are not consistent, however, confusion must result. Direct field labor generally includes welders, electricians, laborers, etc. For engineering work, direct labor includes designers and draftsmen.

Indirect labor: Indirect labor is required to make it possible for direct labor to take place. Indirect labor in the field usually includes supervisors, timekeepers, warehouse attendants, etc. For engineering work, indirect labor includes management, project control personnel, and office staff.

There are several labor categories whose definition as direct or indirect labor frequently causes confusion. These include:

Foreman
Material handling
Testing (e.g., x-ray)
Scaffolding
Purchasing

The project engineer is cautioned to be sure that these, and any other areas of ambiguity, are clearly defined before the work begins.

Contingency: There is so much confusion and ambiguity regarding contingency that it has been given a separate section in this book (see "Estimating Contingency" later in this chapter, and Chapter 10). Suffice to say at this point that contingency may be defined as the provision for those variations from the estimate basis which are likely to occur but which cannot be specifically identified at the time the estimate is prepared.

Escalation: This is the change in price levels over time. For a project cost estimate, the calculation of escalation requires definition of a "base date" which reflects the time period at which the prevailing prices in the estimate existed. The estimator must also define the "centroids" of material, labor, engineering, and other cost categories, which represent the average point in time at which procurement of those items will take place. Escalation "factors" can then be applied to bring the estimated costs up to the pricing level that is expected to prevail at the time of the project (see Figure 5.3) and "Estimating Escalation," later in this chapter). "General escalation" refers to the escalation which exists for the industry in general, whereas "specific escalation" refers to that part of the market which affects the type of project under consideration. Frequently projects experience escalation trends which are markedly different from those of the economy as a whole, due to the effect of supply and demand on the part of the marketplace-affecting projects.

Estimate accuracy: This is another frequently misused term. Estimate accuracy is the tolerance within which there is a specified

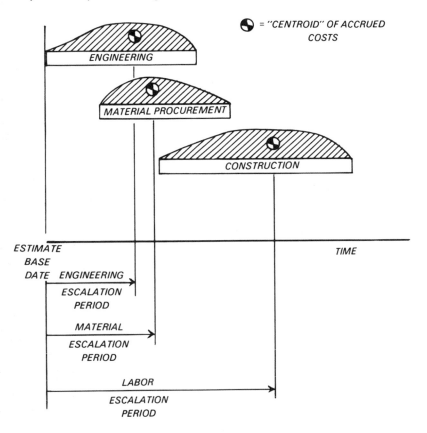

FIGURE 5.3 Calculating escalation for the typical small project.

probability that the actual cost will fall. To express accuracy we need two things: the tolerance band, i.e., the variation above and below the base (usually expressed as "plus or minus X%"), and we must also specify the probability that the final cost will be within the tolerance limits. When asked about the accuracy of an estimate, many estimators will say, "It's plus or minus 10%." This is, unfortunately, absolutely meaningless, as it defines only the tolerance limits. The only correct way to specify estimate accuracy is to say, "The probability is 80% that the final cost will be within plus or minus 10% of the estimate." This is illustrated by Figure 5.4, which shows that, given a tolerance limit, two projects can have very different accuracies. It is, of course, difficult to determine estimate accuracy, and this is discussed in "Estimating Contingency," later in this chapter and Chapter 10.

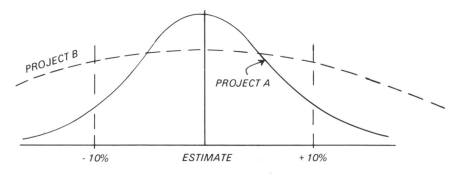

FIGURE 5.4 Defining estimate accuracy. The confidence limits within which there is a specified probability that the actual cost will fall. Project A: 90% probability of being within ±10%. Project B: 60% probability of being within ±10%.

Estimating database: An estimating database is a collection of cost data reflecting a common time period and average past performance. The data is usually organized in tabular form, by cost code, and expressed relative to design variables (e.g., cost per hp) or quantities (e.g., cost per man-hour). Current cost data is used to keep the database up-to-date.

Estimating methodology: An estimating methodology is a set of guidelines, procedures and correlations used to prepare an estimate from the database. While the database provides specific data, the methodology provides generalized relationships between design and cost variables. The database is the basis of the methodology and the methodology is used to prepare estimates.

ESTIMATING MATERIAL COSTS

Material costs are divided into two categories: equipment costs and bulk materials costs.

Equipment Costs

Vendor Quotes

Most companies estimate equipment costs by means of vendor quotes. There are two types of vendor quotes: a budget quotation, which is not binding and which is used solely for budget purposes, and a bid quotation which is binding and which appears on a purchase order.

Although budget quotes from vendors generally have a high degree of credibility, they are not necessarily a good source of estimating data. For instance, the item of equipment for which a budget quote has been given may be quite different from that which is eventually bought. The vendor will usually quote a standard, "off-the-shelf" item which may or may not meet all the specifications which will apply. By the time the design is completed and the specifications included, the actual item of equipment often turns out to be much more costly than that which was quoted originally. Also, the vendor has a lot of incentive to see the project go ahead and the order placed with him. He can, therefore, hardly be expected to be unbiased. In general, he will be inclined to give the lowest budget quote possible, although there may be cases, such as when he is to be the sole bidder for a special item, in which he may want the project engineer to be conditioned to the idea of a high price. In order to compensate for the vendor's bias in budget quotes, a "vendor bias adjustment" is usually added. This adjustment is usually around 10%, but the actual amount will vary depending on conditions.

Even bid quotes need to be conditioned. Most purchase orders start with no revisions but end up with quite a few, most of which result in added cost. Therefore, even a bid quote needs to have a "developmental allowance" added to it, usually in the range of 3 to 8% depending on the complexity of the equipment involved and the likelihood of revisions.

Historical Data

Another way to estimate equipment costs is through the use of historical data. Such data may be collected by the company involved (see "Use of Historical Data," later in this chapter). Or, standard reference manuals are available, as well as computerized estimating services which can be used on a timesharing basis. Many equipment vendors will provide pricing books, formulae, and graphs so that a preferred client can do his own budget quotes. Historical data is often the preferred way to estimate as it reflects actual past experience and includes miscellaneous costs which might otherwise be overlooked.

Checklist

When estimating material costs it is important to recognize the various cost elements that make up the total cost to the project:

The cost of the equipment itself (usually the base quote): This must include all items necessary for complete installation.

Spare parts: these can be "warehouse spares" or "capital spares," which are a high cost, special purpose spare parts (e.g., compressor rotors)

Special packing: some equipment requires special packing which may
not be included in the price

Freight: These changes may vary as equipment may be quoted as
"FOB" (freight-on-boards), "FAS" (freight-along-side), or delivered to site.

Special shipping: This is for complex, fragile, or very large equipment

Vendor reps: These are technical representatives who assist in the
installation and startup of a complex piece of equipment

Taxes: (federal, state, and local)

Duty: (on imported items)

Escalation: is the price fixed or is the price in effect at time of
delivery

Insurance

Exchange rate fluctuations (for imported equipment)

All the above items may contribute to equipment cost. The estimator
should check that all have been accounted for in the equipment cost
or elsewhere in the estimate.

Bulk Material Costs

Bulk materials, such as piping, electricals, structural steel, instrumentation, insulation, and paint, pose one of the most difficult problems in estimating, especially where small projects are concerned. In
most cases, the actual quantities of each type of bulk material will not
be known until long after the estimate is prepared, and, even then,
the many different categories of materials, each of which has a different price, make preparation of an estimate a time-consuming task.
Small projects have the added problem of not having enough time
for detailed estimating, as well as the revamp problem that the bulk-
material quantities may not actually be known with any confidence until
after the construction is complete. This is due to the routing of new
cables, pipes, etc. which, in an existing plant, may be determined by
interference factors which are not evident in the design stage. So
bulk materials are an area of likely overrun, and a realistic estimate is
important to the success of the small project.

Data from Similar Projects

How then to estimate bulk materials for the small project? One way is
to use data from similar past projects. When similar past projects are
used, however, adjustments must be made to reflect the difference between the past project and the current one. For example, we might
note that we used so many tons of piping on a similar project last year,
and that this project should have about 10% more linear feet of piping,
with an average line size 2" greater. A more detailed example is as
follows:

Past Project

Piping cost: $15,000
Piping quantity: 2000 linear ft., 8 in. avg. diam.
Piping complexity: average
Piping materials: carbon steel, schedule 40

Current Project

Piping quantity: 2200 linear ft., 10 in. avg. diam. (price is +21%)
Piping complexity: more flanges and fitting required
Piping materials: carbon steel, schedule 40

Piping cost calculation:
let B = base cost (from previous project)
   Q = ratio of quantities = 2200/2000
   S = ratio of avg. unit price for 10 in. piping/8 in. piping
   C = factor for complexity, more flanges and fittings: say 1.40
   E = escalation factor from then till now, say 1.12

then: piping cost = B × Q × S × C × E
               = (15000)(1.1)(1.21)(1.4)(1.12) = $31,305

Factors Applied to Equipment Cost

Another popular way to estimate bulk materials is with factors, or ratios that indicate the amount of bulk materials in each category that are typically associated with a certain type of equipment. For example, we might multiply the cost of a pump by various percentages to get the cost of the associated piping, electrical, and foundation work. These factors, which can be found in published literature on cost estimating and are used in databanks for timesharing, are generally not well suited for small projects. They reflect average experience, and the specific layout requirements for small, revamp-type projects are apt to be quite different. However, for those small projects which are repetitive, or those for which good data is available (e.g., architectural work), the factoring approach can work well.

Preliminary Takeoffs

A third approach is the use of preliminary "takeoffs," in which the quantities of bulk materials are derived from the measurement of typical distances, taken from drawings. For example, the plot plan might show the distance from some motor-driven equipment to the substation and to the control house. From these dimensions, an average cable-run length can be derived, and an estimate made of the total cable requirements. Even better, more detailed takeoffs may be available, such as those used for purchasing. An advantage of preliminary takeoffs, is that they tend to produce estimates of the right magnitude most of the time. Although less exact than detailed takeoffs, they are less likely to miss major quantities than the factoring method or a detailed takeoff performed too early in the design process. Whenever takeoffs are used, a "takeoff allowance" must be included to cover those items which are likely to have been missed. A typical takeoff allowance can be 25% or more depending on the level of detail of the takeoff.

Once the takeoff is complete, the costs can be estimated by deriving an average unit cost based on past data or on information from vendors and reflecting the mix of sizes and materials that are expected.

ESTIMATING DIRECT LABOR COSTS

Use of the Network and Resource Plan

Like bulk materials, labor costs are apt to be particularly difficult to estimate for small projects. When the network has been completely resourced, we have an ideal basis for estimating the direct labor costs (see Figure 4.1). The total man-hours represented by the resources on the network multiplied by the activity durations can then be multiplied by the appropriate labor rates to obtain the labor cost estimate. This process is straightforward and simple, and can be easily computerized as shown in Figure 5.5.

One advantage of this approach is that the estimator thinks in terms of the numbers of people required to do the work in an activity, rather than the number of man-hours. When estimating is being done by judgement, it is important to define a method which is based on a format for judgement which matches the way people think, and most project people think more easily in terms of "how many people for how long" than in "how many man-hours."

Use of Quotes

For certain types of work, quotes can be obtained from the potential contractor. This practice can be effective in cases where the work involved is quite similar to previous work done by that contractor. However, use of such quotes should reflect consideration of bias, potential growth in the scope of work, and other aspects of the work which might not be reflected in the quote.

Use of Data from Similar Projects

Data from similar past projects can be analyzed and adjusted to derive a labor estimate for the current project. The adjustment process

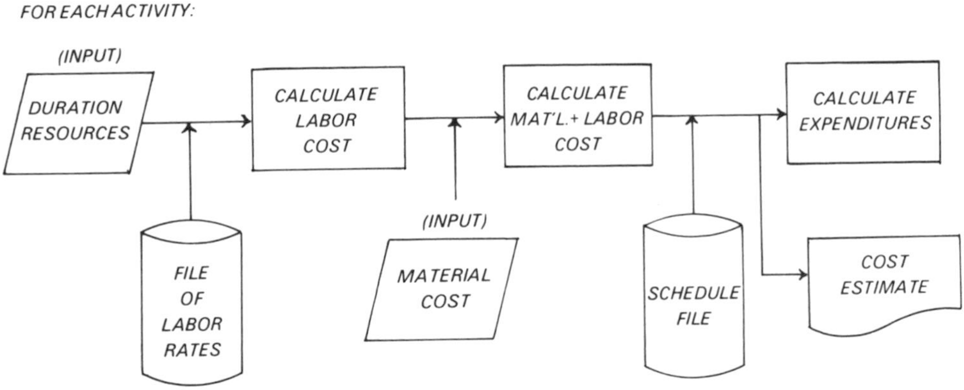

FOR EACH ACTIVITY:

FIGURE 5.5 Computerized estimating of network resources.

should consider the relative complexity and difficulty of the two projects, the quantity of work involved, the type of labor, and the conditions under which it will be done. For example:

Past Project

Labor cost: $20,000
Direct man-hours: 960
Productivity: average

Current Project

Scope of work: greater than past project by 20% (using judgement)
Productivity: expected to be better than the past project due to:

Simpler work: 3% improvement
Summer weather: 2% improvement
Better access: 4% improvement

Wage rate: 15% higher due to escalation
Calculation of labor cost:

($20,000)(1.20)(0.97)(0.98)(0.96)(1.15) = 25,187

Use of Historical Data

Data on unit labor costs (e.g., man-hours per in. or ft. of pipe, man-hours per ton of structural steel, man-hours per cubic yd. of concrete) can be accumulated in a database and used for estimating. This is a preferred method in most cases, because actual data is generally the best predictor of what will happen in the future (if it happened that way on ten similar projects, chances are it will happen that way again). Such unit labor rates are also available in estimating manuals or computer databases which can be accessed by timesharing. When using industry-standard data, however, it must be recognized that the data is representative of some selected locations and conditions, which may or may not reflect the performance of the project in question. Multipliers are generally developed by companies using these outside services to reflect local productivity. Once the man-hours are determined, the hourly rates prevailing at the project location can be applied to get total labor cost.

Use of Factors Applied to Equipment Costs

Labor costs can be estimated in the same fashion as bulk materials, by using factors or percentages of the equipment cost. As in bulk-material estimating, this can be a dangerous practice for small projects as the relationship between equipment costs and labor for equipment and bulk material installation may be quite different for the particular small project than for the average project on which the factors are based.

Components of the Hourly Labor Rate

When estimating man-hours and costs, an hourly labor rate is used, and it is important for the estimator to know what is included and excluded. Hourly labor costs are made up of some or all of the following components:

Direct wages
Taxes (e.g., employer's contribution to F.I.C.A., unemployment tax, etc.)
Holidays
Sick days
Vacation
Overtime premiums
Insurance premiums (for life and health insurance)
Training
Overheads (consisting of supervision, transportation, supplies, support staff, administration, support facilities, and light construction equipment)
Profit (if labor supplied by a contractor)

The estimator should know whether or not these items are included in the hourly rate and, if not, where they are in the estimate.

Productivity

Estimates of labor man-hours must, either explicitly or implicity, contain an assessment of labor productivity, i.e., the number of man-hours required to accomplish a fixed scope of work. Productivity is generally defined as the *planned man-hours expressed as a percentage of actual man-hours required to complete a given scope of work*:

$$\text{Productivity} = \frac{\text{planned man-hours}}{\text{actual man-hours}}$$

Using this definition, productivity over 100% is "good." However, many companies define productivity as the inverse of the above, so the project engineer is advised to establish and stick with a single definition.

Productivity is apt to be different from job to job, even at the same location, so it is important to make any adjustments in the labor man-hours that might be necessary to account for productivity variations for every estimate.

Some of the factors which affect direct labor productivity are:

Job size: Small projects can often achieve levels of productivity that are orders of magnitude greater than large projects in the same location. This is due to their ability to utilize small, well-organized subcontractors, to supervise closely, and to obtain a small, high-caliber workforce.

Job complexity: Small projects involving complex equipment, sophisticated electronics, or unusually difficult construction practices (e.g., excessive work above- and below-grade, extreme congestion, etc.) are likely to experience lower productivity.

Market activity: Periods in which contractors are looking for work are often characterized by high productivity, since the best workers can be made available and contractual incentives for good performance can be negotiated.

Type of workforce; whether union or nonunion: If the data used to estimate man-hours may is based on a union (or a nonunion) workforce, and the job at hand uses nonunion (or union) labor, the productivity may vary.

Type of workforce; whether in-plant or contractor: The plant forces are likely to have different productivities than the contractor forces, and the type of workforce to be used should match the data used in the estimate.

Schedule duration relative to total manhours; for a given project, there is an optimum schedule duration. If we have a tight schedule, increased manpower concentrations will be necessary, with the attendant loss of supervision, increased interference, and reduced productivity. Similarly, a longer-than-optimum schedule can reduce productivity by diluting supervision, as well as by loss of momentum and continuity.

Shift work: Productivity is generally lower on the evening shift than than on the day shift, and lower still at night. These productivity reductions are due to the need for coordination of work between shifts, darkness, and reduced supervision. The evening-shift productivity is perhaps 80% of day-shift productivity, and the night shift is around 60% of day-shift productivity. These figures will, of course, vary greatly between projects depending on working conditions and the effectiveness of the planning and supervision.

Overtime work: Use of overtime generally results in lower productivity due to the problems of maintaining work efficiency over longer periods. There are some cases in which overtime is considered a financial bonus and therefore little extra work is accomplished.

Changes: Field changes can be due to design changes, rework of fabricated equipment or bulk materials, rework of construction work, or a change of the work scope as the project progresses. These changes tend to interrupt the flow of work, and often requires work which is inherently more difficult than the original scope. They therefore tend to reduce productivity.

Based on consideration of all the above, the estimator should make whatever adjustments are necessary to reflect the productivity expected for the project. These adjustments can be judgements, but they should be clearly documented. For example:

Base productivity from reference data:	90%

Adjust for:	project size (smaller than normal)	+5%
	work completely (higher than normal)	(10%)
	market conditions (normal)	n/a
	type of workforce (plant forces)	(10%)
	tight schedule	(8%)
	shift work (20% of work at 15% less productivity)	(3%)
	changes (normal)	n/a

Productivity expected for current project:

$(90\%)(1.05)(0.9)(0.9)(0.92)(0.97) = 68\%$

ESTIMATING FIELD OVERHEADS

Field overheads, which include supervision, indirect labor, temporary facilities, and construction equipment, can be an important part of the small project. Where the work involved is complex, the field overheads can be a major part of the cost. There basically are two ways to estimate field overheads: using a detailed method, or as a percentage of direct labor.

Using a Detailed Method

The detailed method is preferred for many small projects because it enables the estimator to use the particular requirements for the project at hand, which may vary substantially from previous projects. For example, a small project may require a large mobile crane to install a certain item of equipment. The cost of this crane might represent a significant percentage of the project cost, and an abnormally high percentage of direct labor. Small, complex projects might also require an abnormal amount of supervision. The detailed method, then, requires the estimator to specify and price each category of field overheads. One of the best ways to do this is by using the planning network and loading the resources for field indirects (i.e., indirect field costs) onto each activity.

The principle of resource planning is that *the resources required to accomplish the given scope of work in the given duration must be provided if the work is to be accomplished.* These resources can include major items of field overheads such as construction equipment. The same resource-loading tables developed in Chapter 4 can be used (see Figure 4.1). Other field overheads, which are not assignable to specific activities, can be estimated in detail or as a percentage of direct labor. The field overheads on a small project may include:

Indirect labor: This includes the timekeeper, accountant, payroll clerk, guard, purchasing agent, warehouseman, equipment operator, driver, x-ray technician, surveyor, mechanic, clerical and office staff, site cleanup crew, and janitor.

Supervision: Including the foreman, planner, inspector, site manager, construction superintendent, safety supervisor, craft supervisor, field engineer, and subcontract administrator.

Temporary construction: This may include the field office, warehouse for construction material, equipment shed, pipe fabrication shop, other fabrication facilities, change houses, field toilets, guard house, cafeteria, first-aid facility, and laboratory. The furnishings and equipment for these temporary buildings must also be considered as well as the temporary power, water and sewer lines, and phone connections. Roads, ditches, parking areas,

staging areas, fences, and signs are also temporary facilities
which may be required. Note that all the above refers only to
temporary construction, i.e., that which will be removed at the
end of the job. Permanent facilities are part of the workscope
and are not included in this account.

Consumables: Including welding rod, gloves, protective clothing,
small supplies (e.g., nuts, bolts, gaskets, nails, and clamps),
lubricants, fuels for equipment, hardhats, ropes, rags, etc.

Construction equipment: Including lifting equipment (e.g., cranes,
gin poles, derricks, and hoists) automative (e.g., cars, and
trucks of various types), earthmoving, (e.g., backhoe, bulldozer,
front-end loader, roller, and excavator), cranes (e.g., "cherry-
picker" mobile crane, crawler cranes, and sideboom tractor), welding equipment (e.g., welding sets and stress-relieving kits), shop
equipment, testing equipment, communications (e.g., radio and
telemetry), special-purpose equipment, and miscellaneous small
equipment (e.g., pumps, stirrers, hammers, compressors, blow-
ers, ladders, scaffolding, etc.).

Calculating Field Overheads as a Percentage of Direct Labor

This approach is often used for large projects, in which the field indirects are fairly consistent. It is appropriate for small projects which
are reasonably consistent in terms of size and complexity. Since certain field overheads are required regardless of job size, the overhead
percentage can be highly variable, and the estimator is cautioned to
be sure that the data on which the percentage is based is consistent
with the project at hand. If not, an adjustment can be derived from
past data by developing a curve of field overhead percentage as a
function of total man-hours.

INTEGRATED COST ESTIMATING

Many companies have, in recent years, formalized the process of estimating in a way that integrates cost, time, and resources. In fact,
this practice has become sufficiently common that the expression "C-
T-R" has gained acceptance as a way of describing cost, time, and
resource data.

The C-T-R approach to estimating begins with the project plan as
shown on the management network. For each activity, the resources

required and the total cost are estimated on a C-T-R sheet. (see Figure 2.6). The entire estimate is then contained in the CTR catalog, which contains a C-T-R sheet for each activity on the network.

The C-T-R format is a marked departure from traditional ways of preparing and presenting estimates. It has found rapid acceptance because it clearly shows the relationship between cost, time, and resources, and is therefore more flexible than traditional methods and better able to show how costs are affected by variations in the project execution. The C-T-R format is particularly attractive for many small-project applications because of its simplicity, logic, and flexibility. And, since many project engineers are not highly-trained estimators, the change to a different method is not difficult. The C-T-R also provides an excellent basis for project control, see Section III.

ESTIMATING ENGINEERING COSTS

Engineering costs are estimated in a manner similar to that used to estimate direct labor costs. Acceptable methods use the resourced network plan, quotes from engineering contractors, data from similar projects, and historical data.

Use of the Network and Resource Plan

If the network has been resourced for engineering work, we have an excellent basis for estimating engineering costs. Using the same procedure as for labor costs, the number of people in each category × the duration of the activity × the appropriate man-hour rate will give the cost estimate, as illustrated in Figure 5.5.

Use of Quotes

Budget quotes for engineering work are difficult to obtain, since most engineering work is done on a reimbursable cost basis. However, in small projects where the amount of engineering work is readily defined, it may be possible to get a lump-sum budget quote.

Use of Data from Similar Projects

Data from past projects is often a good indicator of the engineering content of a project. The data should, of course, be adjusted to reflect the scope and complexity of the project at hand. This adjustment is often made by calculating the engineering costs as a percentage of direct material and labor cost.

Detailed Estimate Based on Scope of Work

Engineering work can be estimated in detail by estimating the number of drawings and specifications required, based on the number of equipment items and piping lines contained in the basic design. This approach is not generally suited to the small project, as the necessary data and methods are usually not available. However, in some cases it is possible to calculate the number of drawings and specifications and multiply by the estimated hours per drawing to yield the total design manhours. Other types of man-hours can then be estimated in detail or calculated as a percentage of the design work.

Components of the Hourly Engineering Rate

As engineering costs are usually handled in terms of man-hours and cost per man-hour, it is necessary to know what is included and what might be excluded in the rate being used. Those necessary items which have been excluded should be covered elsewhere in the estimate. Hourly rates for engineering are made up of some or all of the following components:

Salary
Taxes (e.g., employer's contribution to F.I.C.A., unemployment tax, etc.)
Holidays
Sick days
Vacation
Overtime premiums
Insurance premiums (e.g., life and health insurance)
Recruiting
Training
Overheads (e.g., supervision, office facilities (floor space, utilities, furnishings, etc.), computer facilities, office staff and services (secreterial, mail, telex, telephone, janitorial, and cafeteria), marketing expenses, cost of financing, company liability insurance, recruiting, internal R & D, library and databank, specifications, procedures, etc.)
Profit (if engineering work provided by a contractor)

Productivity

Engineering, being a form of labor, will experience productivity variations under differing conditions. As in labor productivity, engineering work is done far more efficiently during periods of low contractor activity. During these periods, contractors are likely to cut back on staff—so that the best people are available to work on the project—and more favorable contract conditions can be negotiated to help assure good performance.

Job size and complexity will also affect engineering productivity, with smaller and simpler projects capable of higher productivity. However, in an engineering operation, the small project is apt to get far less attention, perhaps even lower priority, than a large one and therefore the productivity may well be less.

Schedule and job size affect engineering productivity in the same way as they do field labor. Tight schedules foster low productivity as progress becomes more important than efficiency. In addition, the use of overtime often leads to significantly higher costs and lower productivity. Engineering work, particularly on a small project, is apt to be affected strongly by the quality and characteristics of the people involved. One should make an extra effort to identify those individuals who are most likely to do a good job, and have them assigned to the project. (This is also true for key personnel on the field supervision staff.)

ESTIMATING MISCELLANEOUS ITEMS

No estimate is complete without consideration of the various miscellaneous items which affect all projects. Typical miscellaneous items include:

Taxes (federal, state, and local)
Fees (for special operations such as large transport over city
   streets)
Insurance (for materials, personnel, liability, etc.)
Permits
Interest charges
Exchange rate fluctuations
Legal assistance

These miscellaneous costs can be estimated in whatever way is most suitable. The important thing is to be sure all miscellaneous costs are covered in the estimate.

ESTIMATING ESCALATION

Most small projects involve a relatively short time period, so escalation is not as big a problem as it is for larger projects which are conducted, in some cases, over a decade or more. However, the elapsed time from the base date of the estimate (i.e., the time period represented by the data used for estimating) to the time at which the work will actually be done can be significant. Most small project cost estimates must, therefore, make some provision for escalation.

Estimate Components

The calculation of escalation costs usually begins by breaking the estimate into components which are characterized by distinct escalation rates. The estimate components which tend to have distinct escalation trends are:

> Direct materials (equipment and bulk materials)
> Direct labor
> Field overheads
> Engineering services
> All other costs

Escalation Indices

An escalation rate is then established for each cost category. This rate may be calculated a number of ways but it must be remembered that we are looking for the specific escalation that relates to the project, not escalation rates for consumer or other unrelated goods. The sources of data and indices for tracking and forecasting escalation include:

> Federal *government indices* (e.g., Bureau of Labor statistics)
> Escalation indices developed by *industry groups* (e.g., labor unions)
> Escalation indices published in *professional journals* (e.g., Oil and
>   Gas Journal's "Nelson Cost Index")
> *Historical data* from company files or database
> *Economic reports* by leading banks, universities, and industry
>   groups (e.g., The Wharton econometric model)
> Guidelines and forecasts published by the *financial department* of the
>   project engineer's company (used for forecasting, budgeting,
>   financial planning, etc.)
> Services provided by *consulting firms* to industry on a time-sharing
>   basis (e.g., Chase Econometrics)

For small projects, it is best to develop and agree which indices are to be used, update the indices periodically, and use them on all projects.

Calculating Escalation

To calculate the escalation costs, the project schedule is used to establish the centroid of the expenditure of each category of project cost, i.e., the point in time at which, if we purchased everything in that cost category at once, our cost would be the same as the actual condition in which we make purchases over time (see Figure 5.3). The length of time from the base estimate date to the centroid is the time period for which escalation is to be calculated. This amount of time, multiplied by the escalation rate per unit of time, gives the cost of escalation for that category. If the estimate is to be used for cost control, the escalation must be factored into each item in the estimate, so that the escalation account disappears and each cost item is shown as "money of the day," that is, the expected cost at the time the money is spent.

ESTIMATING CONTINGENCY

Estimates Must Include Contingency

One item that is, by definition, included in all estimates is uncertainty. Since an estimate is, as we have said, a prediction, it is inevitable that the project actually develops in many different ways from what we expected. In most cases, it will turn out to be different in a way that results in increased cost. To bring the cost estimate up to a level which has a reasonable chance of not overrunning, we add contingency.

Contingency is a much misunderstood cost engineering concept. It is also the part of cost estimating for which many unsatisfactory estimating methods exist. Small projects have a risk of overrunning—in many cases the risk is greater than for large projects—so contingency must be included in the estimate to cover design changes, variations to the planning basis, and estimating differences with actual costs. Chapter 10 is devoted to this subject and describes a number of methods for estimating contingency. For convenience, some of the methods are summarized below.

Contingency calculations are particularly important for cost estimates that are to be used for bidding purposes, in which we wish to study the relationships between the bid price and the probability of winning the job.

Summary of Methods for Estimating Contingency

Judgement, Based on the Risks Foreseen for
the Project at Hand

This commonly used method can actually provide reasonable results in the small project environment, but it suffers from being purely qualitative. Studies have shown that engineers and managers tend to be unduly optimistic about the chances for success. And, there may be risks which are overlooked in such an analysis. Sometimes, contingency assessments are made by committee which, depending on the group dynamics involved, may give a better (or at least a more acceptable) result.

Guidelines or Procedures Derived from
Historical Data

Many companies set contingency by policy, such that all estimates submitted for approval are given the same contingency. This method,
probably the most commonly used, ignores the fact that some projects
have more uncertainty or risk than others, and therefore require more
contingency. Those high-risk projects are likely to have contingency
"buried" in the estimate, thereby fostering a sense of confusion and
mistrust, as well as diminishing the estimate's usefulness as a control
tool. Another problem with this approach is that historical data, i.e.,
what has happened, is not always the best predictor of what will
happen.

Comparison with Past Projects Having
Similar Risks

This method can give reasonable results for the small project, if the
selection of projects for analysis and comparison is made carefully, and
the analysis done objectively. For example, the project engineer in
this case might say "We had 20% contingency on a similar project last
year, and needed every bit of it. However, the current design reflects certain lessons learned from that project, and so the portion of
contingency that covered those unforeseen changes will not be needed
on this project and we can use a contingency of 17%. This project also
is in a more difficult area of the plant to work in, but we had better
data to use in our estimate." This thought process can be documented
as follows:

---

contingency provided on similar past project: 20%
final cost as % original estimate: 123%
portion of overrun due to contingency variations: 1%
thus, contingency used on past project: 21%

differences in estimating bases between past and present projects

design: current project has less potential for changes
planning: current project is in a more congested work area
cost: current estimate used better data

contingency required for current project

design variations: say 3% less than past project
planning variations: say 5% more than past project
estimating variations: say 6% less than past project

  net requirement = 21% − 3% + 5% − 6% = 17%

Use of a "*50/50 Project Model*"

In this and the preceding chapter, we have seen how a small project is planned, scheduled, resourced and estimated using the integrated technique which results in the definition of the "project model." The original project model is comparable to the base estimate in that it contains no contingency for cost or schedule variations. Therefore, due to the inherent skewness in project data described previously, the base model is more like a "target," that is, it represents a cost and schedule performance that is worth trying for but which we do not expect to achieve. So, the probability of the "target model" being achieved or bettered is less than 50% and probably about 25%.

It is possible, then, to create a "50/50 model" in which the durations and costs reflect the variations covered by contingency, and the probability that the cost and schedule will be met or bettered is 50%. This model can be compared with the target model and, the differences considered to be contingency.

This approach is well-suited to small projects, as it shows clearly the cost and schedule effects of the expected variations, it is easy to do, it avoids use of statistical concepts and terminology with which (unfortunately) few people are comfortable, and it can be easily related to past data.

Manual Calculations Using Probabilities

A number of methods are available for calculating contingency by identifying specific risks, assessing their probability and cost impact, and performing probabilistic calculations. The best known methods in this category are "decision tree" analysis, which uses a logic diagram, and "expected value" calculations which use a tabular presentation. Although one or both of these methods can be useful in certain situations, because they deal with discrete risks they are better suited for risk analysis and decision-making than for contingency calculations. There are a number of excellent texts on this subject for the reader who wishes to try these methods.

Probabilistic Simulation

There are a number of computer programs available which do a credible job of calculating contingency using a technique known as "Monte Carlo Simulation." These programs may be suitable for some small projects in which a detailed and rigorous approach to risk analysis and contingency calculation is required. The simulation technique attempts to define the probability function for the project cost by simulating a large number of possible cost outcomes, using, as input, probability functions for each of the cost variables. This technique is widely used in many other applications, and is well-accepted. Its disadvantage in the project environment is that the cost variables must be independent for the method to work properly, and, in most cases, this is not the case. The reader desiring more information on this technique is again referred to the many excellent books on risk analysis, operations research and decision theory which describe it thoroughly.

ESTIMATE PRESENTATION

Once the estimate has been completed we are ready to summarize and prepare it for presentation and review. This part of the estimating process often proves to be just as important as the estimate itself, for, if the estimate lacks credibility, it will not be accepted by management, and the project may not be approved. If this happens, the estimate will be useless as a control tool. How then can an estimate be documented and presented so as to assure it's credibility and usefulness?

Defining the Estimate Basis

The most important aspect of estimate documentation is not the estimate itself, but it's basis. The "tripod" of the design, planning, and cost bases is what supports the estimate, and it can be no better than that base. Thus, the estimate basis must be clearly defined:

The design basis: This includes specific drawings, sketches, and specifications; date of release, revision number, and person(s) responsible; important design assumptions; overall scope of work understood to be involved; specific inclusions and exclusions of facilities to be installed; and work to be done.

The planning basis: This involves the network plan and schedule, resource plan, and contracting plan; date of release, revision number, and person(s) responsible; and important assumptions as to how the work will be executed (e.g., hours worked per week, use of overtime, contractor performance expected, etc.).

The cost basis: This involves the estimating method(s) used (and why they are appropriate for the project); sources of estimating data; cost level of the base estimate (i.e., what point in time is reflected); productivity and hourly cost rates used for engineering and labor; and escalation rates used.

The Estimate Summary

The estimate summary should provide a one-page summary of the costs, generally organized as follows:

Direct material
  equipment
  bulk materials

Direct labor
Total direct costs
Field overheads
 supervision
 indirect labor
 construction equipment
 temporary facilities
Engineering
 in-house
 contracted
Miscellaneous
Escalation (if not included above)
Total base estimate cost
Contingency
Total cost estimate

The estimate summary can then be further broken down by areas,
cost centers, or according to any other practice followed by the company.

A final small but important point: cost figures should be rounded;
nothing makes a worse impression than precise estimate figures (i.e.,
to the nearest dollar or penny) which imply an estimating accuracy
that clearly is not possible.

Reconciliation to Past Estimates

The reconciliation is a very important aspect of estimate presentation
which is often overlooked. Just about every estimate has some sort of
estimate which preceeded it, and every manager, looking at the new
estimate, will mentally compare it with the old. Credibility is characterized by clarity and thoroughness (but without excessive detail),
as well as by having the answers to unasked questions. The reconciliation provides all these characteristics. The reconciliation should
be rigorous but easy to understand. Large reconciliation items can
be explained with footnotes as needed. The method for preparing reconciliation is described in "Reconciling Forecasts and Estimates,"
earlier in this chapter.

Estimate Sensitivities

In some companies, it is desirable to present to management the specific
areas which the estimator feels are particularly likely to cause an overrun. In many cases, the credibility of the estimate is enhanced by an
open, frank, and well-thought out presentation of the specific risks
and sensitivities involved (note that these are specific risks which are
not included in the contingency analysis). For this type of manager,

a page showing the sensitivity of the estimate to various risks is often appreciated. For example:

Total cost estimate	$35,000
Possible scope change (new road)	+5,500
Possible scope change (rework stack)	+1,500
Possible strike (assume three weeks)	+8,500
Overseas purchasing (compressor)	(2,000)

It is tempting to add up to the plusses and minuses to get an idea of the total possible cost. This should never be done, as the laws of probability indicate that the probability of all these things happening together is the product of each individual probability and is therefore almost always infinitesimally small (see Chapter 3).

Estimate Details

The estimate details are unlikely to be reviewed by a manager, but they are important for building credibility with those who are doing the work (the project engineer, the contractor, and others). The format for the estimate details varies widely from company to company. What is important is that the details be clear, explicit, and appropriate for their intended use (i.e., possess "quality").

The intended use of the estimate details is cost control. Therefore, the level of detail in the estimate summary should be comparable to that of those who will be using the estimate. Unfortunately, many good estimates turn out to be useless for cost control because they have such an excess of detail (usually provided out of a misplaced desire to impress) that the work required to extract meaningful information is not worth the benefit.

Some estimates which have used adjustments from similar projects or other gross estimating techniques contain too little detail for control and this should be recognized before attempting to use them for that purpose. In those cases, detailed data from past projects can be used to establish a breakdown of costs for the control estimate.

REVIEWING AN ESTIMATE PREPARED
BY OTHERS

Project engineers, managers, and cost engineers often find themselves in the position of reviewing and either accepting or rejecting an estimate prepared by someone else. The estimate may have been prepared by a subordinate, by a contractor, by the estimating department, or by another project engineer. Those who have had to review estimates have probably found that it is not very easy. What questions should I ask? What should I look for? How can I be sure that the estimate is as good as one I did myself? These questions are important because the estimate reviewer will probably have to approve it, and therefore share in the responsibility for it's being correct. This problem of estimate reviews is often made much worse by the fact that the time left for reviewing is usually very short by the time the estimate is finished, and no one wants to be accused of holding up the project.

Objectives of the Estimate Review

Fortunately, there are some guidelines which can be followed. Some are derived from the estimating fundamentals described in this chapter, and others are borrowed from our friends the auditors.

An estimate review can be thought of as a series of questions and checks, designed to accomplish the following reviews and answer the following typical questions:

1. *Evaluating the basis of the estimate*: It is complete and is it likely to change? Is there any new technology involved? If so, how was that reflected in the estimate? Are the plan and schedule well thought-out? Is the schedule a 50/50 case? If not, have schedule premiums been included? Are there are unusual aspects to the construction work? If so, how have they been reflected? How is this project similar to previous projects? How is it different? Have the differences been accounted for? Are the assumptions on which the estimate is based clearly stated and reasonable?

2. *Reviewing the methods and data used*: Are the estimating methods appropriate to a project at this stage? Have these methods given good results on previous projects? What parts of the project were difficult to estimate? How were they handled? What was the source of estimating data? Is it representative of conditions anticipated for this project? Does it have the correct amount of detail? Has it been adjusted where necessary to reflect this project? Is it current? Does it exclude any elements which should be in contingency? Was a computer used for estimating? If so, how was it used? Was the output verified?

3. *Assess the caliber of the personnel involved*: Are they experienced in estimating? Are they aware of the parts of the estimate most subject to variation? Have they spent enought time and effort on the critical estimating variables? Did they check their work? Was it properly reviewed?

4. *Evaluate the quality of documentation*: Can every number on the summary pages be traced back to the backup? Is it clear how each number was calculated? Is the estimate in a format suitable for cost

control? Is it clear that the estimate has not been "padded" in any way? Have all assumptions been clearly documented?

5. *Check the absolute value of the estimate*: Is the estimated cost consistent with other similar projects? Are the relative values of the major cost components about what one would expect? What special features does this project have which might make it difficult to base the estimate on past experience? Is there any reason to expect that the estimate is biased high or low?

Techniques for Estimate Review

The techniques for estimate review are based on these principles:

Avoid Getting Involved in the Details Unless Absolutely Necessary

Many of the mistakes made by estimators and project managers happen because they are so swamped by details that it becomes impossible to draw any conclusions. Many project engineers find themselves so busy examining the "bark on the trees," that they never get to realize that the "forest" is blazing away behind them. Those who review an estimate by checking the arithmetic (not an uncommon practice) are good examples of this practice which can be improved by refusing to get involved in details unless absolutely necessary. Be aware that this may be difficult, as the "reviewee" may feel safe by keeping the reviewer bogged down in detail.

Remember the "80/20 Rule"

This is also a favorite of auditors and consultants, also called "Pareto's Law," the "ABC Rule," and the "Law of the Significant Few and Trivial Many." It simply states that there are, in any situation, a few items or aspects (say 20%) which make most (say 80%) of the difference. This is definitely true for cost estimates. Almost inevitably, the cost variables which have the greatest impact on project cost, and which vary the most are:

Labor productivity
Bulk material quantities
Unit man-hours for labor
Engineering manhours
Hourly rates for labor and engineering
The scope of work
Escalation (if applicable)
Design changes
Market conditions
Inclusions and exclusions
Contingency

In general, if these aspects of the estimate have been handled correctly, there is not a lot that can be wrong with it. So, it is wise to spend most of the time for estimate review on these items which, if wrong, have the greatest impact.

Use the Auditor's Questioning Method

Auditors are skilled at asking questions which will result in an answer that provides a good indication of whether more questions need to be asked. For example, a satisfactory answer to question 1 means that we can proceed to question 2 whereas an unsatisfactory answer means that there may be some weakness in that area and we better explore it more deeply (with question 1a). Skilled managers who seem to have the uncanny ability ability to, in five minutes, pinpoint the one mistake in a piece of work that took weeks to do, are probably using this technique. This approach works very well in estimate reviews, since there clearly isn't time to go through the entire estimate. It depends, of course, on the questioner's ability to be unimpressed by charm and by being told what one supposedly wants to hear.

Use the Auditor's Sampling Method

Auditors also like to sample; to take an item at random and thoroughly check everything there is to check about it. If everything holds up well, chances are the other parts of the estimate are reliable also.

Use the "Vertical Audit" Technique

This technique is used most often in conjunction with sampling and questioning. It involves following, in detail, a procedure which is typical of the work done. For an estimate review, it means examining every step of the estimating process. The following conversations illustrate the concepts discussed so far. They might take place between the estimate reviewer and the person responsible for the estimate:

Q: "Where did these electrical costs come from?"
A: "The electrical department. Those guys really know their stuff. In 30 years, they've never been wrong on an estimate. You can be sure those electrical costs will be right on the money!"
Q: "How was the estimate done?"
A: "The same way they do all their estimates. Why should they change when they've been so successful?"
Q: "How was it documented? Where is the backup to the estimate? Where are the quantities? How were they derived? On what drawings were they based? Were those drawings up-to-date with the latest changes? What is the basis of the cost data? Is it relevant to this project? Is it consistent with the rest of the estimate?

What have you done to make sure that it includes everything that
should be included? Who did the electrical estimate? Who checked
it? When was it done?"

A: "Uh, we better go talk to the electrical department."

In the electrical department, of course, a similar conversation takes
place, going, if necessary, into progressively greater levels of detail
until the reviewer is satisfied that the estimate is sound or has con-
cluded that it is unacceptable. Compare that conversation with this
one:

Q: "Where did these electrical costs come from?"

A: "The electrical costs were prepared by the electrical department
as they have the special knowledge and up-to-date data which is
needed."

Q: "How was the estimate done?"

A: "The scope of work to be estimated by the electrical department
was defined by the estimating department, along with the method
to be used. The design quantities were derived from the current
single-line drawings, equipment specifications, and plot plans.
The cost data was taken from our current purchasing agreements
with suppliers of cable, conduit, and fittings, as well as from
budget quotes from two potential suppliers of the transformers
and switchgear."

Q: "How was it documented?"

A: "The electrical estimate was documented in this package which
was transmitted to estimating, reviewed for consistency, and then
incorporated in the project estimate."

Q: "Who is responsible for the Electrical Department's work?"

A: "Bob Kable."

Q: "OK, let's call in Bob Kable and we'll all have a look at this
package."

Note that the confident reviewee in the second conversation may only
be telling us what we want to hear (though it sure sounds good), so
we must make absolutely sure by insisting on going through the written
details anyway. By involving Bob Kable, we make sure that we get an
assessment of the people who actually did that part of the work. In
any case, if the electrical estimate turns out to be sound, we can go
through the rest of the project estimate with a more cursory review.

Use a Quick, Independent Check Estimate

There is no better way to build confidence in an estimate than to be
able to arrive at the same answer in a different way. Even if one is
reviewing a detailed estimate, it is almost always possible to check it by

coming up with a "quick and dirty" check estimate. Techniques for doing the check estimate include adjustments to similar past projects, rule-of-thumb estimates, and curve-type estimates. If one can show that the estimate being reviewed is "in the same ballpark" as a rough, independent estimate, chances are it is reasonable, If not, of course the check estimate may also be wrong, but it indicates that some time should be taken for the estimate review.

EXPENDITURE FORECASTING

As part of the budget package the project engineer generally must prepare an expenditure forecast, to enable the financial department to assure that the necessary funds will be available when needed. This part of the estimating process is notorious for being difficult and un-satisfying as it tends to be a time-consuming task which generally produces erroneous results. The integrated approach to planning, scheduling, resourcing, and estimating provides us with some better ways to forecast expenditures. In this context, we use the term "ex-penditure" to mean the actual passing of funds from one company to another. The "expenditure forecast" shows how much will be expended each week or month.

It should be noted that many companies forecast expenditures by using standard curves similar to S curves. However, just as the tra-ditional S curve is a poor predictor of small-project man-hour and prog-ress trends (see "Details of Resource Planning," in Chapter 4), so too it is apt to falsely predict small-project expenditures. The tech-niques described below are easy to implement and will generally pro-duce better results.

General Principles of Expenditure Forecasting

Step 1: Calculate the Expenditure Timing

Expenditure calculations must be done recognizing that the expenditure pattern will be different for different types of costs:

Labor costs (engineering, direct and indirect labor): If the work is done on a reimbursable basis, an invoice is usually submitted month-ly for work during the previous month so labor expenditures tend to be linear over time.

Overhead costs (field and office overheads): These costs will also generally follow a linear expenditure pattern over time. They may be directly reimbursed, or included in the hourly rate.

Material costs: Materials are usually invoiced when delivered, fa-bricated materials may require progress payments. The expenditure pattern for materials tends to be more like a point or series of points.

In all cases, it is reasonable to assume that there is an elapsed time between the receipt of an invoice and the payment. This is due to the normal corporate practice of checking and approving the invoices, as well as holding on to company funds as long as possible. In most cases, a 30-day period is to be expected, but the accounting department is the best source for information on each company's practices. This lag in payment should be allowed for in the forecast. A general formula for calculating the timing of an expenditure is:

Date work is completed + time to prepare and submit invoice + time to pay

The project's schedule is, of course, the primary tool for determining the timing of the major payments.

Step 2: Calculate the Expenditure Amounts

The estimate and the contract plan are the basis for calculating expenditure amounts. The estimate allows us to group costs into major contracts (if this has not already been done). The type of contract will then determine the specific timing:

Lump sum: This kind of contract may gave progress payments, or may be paid in full at the end of the work.
Reimbursable: These may be owner-financed (in which case payments are made in advance) or contractor-financed (payments made monthly for work performed). May have a retention (usually about 10%).
Bonus: This may be included and is usually paid at the end of work.

Using the Integrated Network for Expenditure Forecasting

Having prepared a resource- and cost-loaded network we know, from the schedule, when each activity will take place and how much each will cost. It is therefore straightforward to calculate how large the expenditures will be and when they will be required.

Resource aggregation provides an indication of the man-hours spent in each time period. If the estimate has been done on this basis we also know the cost of the invoice that will be submitted. By inserting the timing factor, we can easily generate the expenditure forecast. If the basic plan and cost estimate have been done with a computer-assisted method, the expenditure forecast can also be easily automated.

Using the Escalation Calculation to
Forecast Expenditures

Since the escalation calculation requires us to segregate the estimate
into categories for escalation which are similar to those for expenditure forecasting, and to locate those cost centroids in time, we can
also use that calculation for expenditure forecasting. This is the process of "spreading" the escalated costs about the centroids used for
escalation.

APPROXIMATE ESTIMATING

Project engineers frequently encounter the situation in which a cost
estimate is required immediately, must be prepared for a project about
which little is known, and must be prepared in the absence of estimating data and methods. In other words, a situation requiring an approximate estimate. Approximate estimates, though not very accurate,
have a number of important uses:

As a basis for decision making very early in a project's life
As a check on an estimate or bid prepared by someone else
As an interim estimate to justify further funding and studies

There are two basic techniques for approximate estimating:

Adjustments to past projects
The "sixth-tenths rule"

Each of these is discussed below.

Estimating by Adjustments to Past Projects

This method uses the same technique described for reconciliations and
in the estimating of bulk materials; that is, one should start with a known
cost data-point, and make adjustments to reflect differences between
the past project and the current project in terms of the design, planning, and cost basis. In some cases, it may be appropriate to make
these adjustments using the six-tenths rule which is discussed in the
next section. The technique is best illustrated by the example below.

Known cost of past project:		$100,000
Adjust for:		
Design variations		+$2,000
equipment sizing	+$3,000	
bulk materials, size and quantity	+$2,000	
type of materials	($3,000)	

Planning variations		($1,000)
schedule duration	+$4,000	
type of contract	($2,000)	
shift work	($3,000)	
Cost variations		+$12,000
escalation	+$12,000	
Cost of Current Project:		$113,000

In preparing approximate estimates it is important to clearly document the judgements, assumptions, and calculations made, no matter how approximate they are. In this way, one can always explain how the numbers were derived and, if the guesswork is criticized, the critic can always be invited to provide a better guess!

The Six-Tenths Rule

Cost engineers have observed over the years that, in many cases, the cost of something does not increase proportionately with its capacity. For example, if we wish to build a storage tank with twice the capacity of an existing tank, we will probably find that the bigger tank will not cost twice as much. Many people refer to this phenomenon as *"economy of scale."* In this case, we can easily deduce why economies of scale are in evidence. If the volume is doubled, the surface area of the tank will only increase 40%, and, since the surface area determines the weight which in turn determines the cost, the cost is likely to increase by only about 40% as well. For more complex facilities, there are many cost components that do not increase in a one-for-one manner with capacity. For example, instrumentation costs do not, in general, double when the size of the equipment is doubled.

When cost vs. capacity data is plotted, if often tends to take the form:

$$\frac{\text{cost of A}}{\text{cost of B}} = \left(\frac{\text{capacity of A}}{\text{capacity of B}}\right)^n$$

where n is typically about 0.6 (although it will vary).

This relationship appears, handily, as a straight line on logarithmic paper.

Different cost categories will have different values of n (found in various references), and the reader is cautioned not to extrapolate too far using the above relationship. However, the six-tenths rule is a handy way to make adjustments for approximate estimating.

CHAPTER SUMMARY

In this chapter we discussed the basic concepts and techniques of cost estimating, and demonstrated various ways to apply them to small projects. Having taken our hypothetical small project through the steps of network planning, scheduling, resource planning, and cost estimating we now have a complete "budget plan." As we seek project approval, we can advise management about exactly what we intend to do (the design basis), how we intend to do it (the resource and contracting plans), how much it will cost (the estimate), and when we'll need the money (expenditure forecast). That is certainly an acceptable basis for the important process of project approval.

But the budget plan is a lot more important than just a means of getting the project approved. It serves as a basis for communication, commitment, and control—clear communication on our part as to what we understand our job to be and how we intend to do it, firm commitments on the part of those who must supply the resources we need, and effective control of our own work and that of others as the project progresses.

In Section III, we will see how much of our project-control capability is based on this budget plan.

The budget plan is also a fully integrated project model which we have already used, during its formation, to optimize resources, time, and costs. Once approved, it becomes the "static model": to the benchmark against which we measure what actually occurs.

6
Organizing for Small Project Management

THE EFFECT OF THE ORGANIZATION ON
SMALL PROJECTS

In Chapter 1, project-management problems related to small projects
are described. Most companies have some or all of the conditions described in that chapter which include:

Priorities on the allocation of resources generally favor production
Lack of formal methods, procedures, systems, and reference data
for use in planning and estimating
A lack of authority to match responsibilities of the project engineer
Unclear lines of communication and responsibility
A lack of appreciation of the small project's difficulties due to their
revamp aspects, the fact that many projects are managed at once,
and the difficulties of working in an operating plant

It is evident from the above list that many of the problems of small
projects are caused internally and therefore can be corrected internally. This chapter is written for project engineers or managers who
have an interest in and an ability to change how things are done within
the company. It discusses ways to strengthen the project-engineering
function, as well as ways to remove obstacles to the effective performance of that function.

DEFINITION AND COMMUNICATION OF MANAGEMENT OBJECTIVES

General Objectives for Small Projects

One of the most frequent causes of conflict between project engineers and those who operate the plant facilities is the different interpretations of management objectives. Although maximizing profit is the clear overall objective, the way in which engineering and operations functions each contribute to profit is often not clear. In the absence of guidance from plant management, the obvious contribution of production to profit, when compared with the overhead cost of engineering functions, tends to make any task that does not directly contribute to production appear a low priority.

This problem is, of course, nothing new. It is manifested most often in the allocations of plant resources. Since the timely application of the required resources is essential for projects to be completed on schedule and within budget, this problem can have a profound effect on the ability of the project engineer to do his job.

The way in which small projects contribute to profit can be clarified by defining specific management objectives for each project or group of similar projects. Where necessary, guidelines should be developed for making decisions on resource allocations between projects as well as between project work and operations work.

Guidelines for Decision Making

On every project, large or small, the focus of the project engineer is on schedule, capital cost, and design quality (not necessarily in that order). As we saw in Chapter 2, these parameters are directly related, and it is generally not possible to implement an improvement in one without affecting the others in a negative way. As a result, project-management decisions trade off one parameter against the other until the optimum combination is found. For example, a proposed design change adds quality by improving the maintainability of the unit. There is an investment cost to implement the change, as well as a schedule penalty. Should it be done? The answer lies in the relative importance to the company of maintainability, cost, and schedule. This relative importance depends on considerations of profitability which are by no means obvious. So, even in this simple and highly typical example, it is evident that, without some management guidelines, there is no way that a project engineer or manager could make a decision and be sure that he had acted in the company's best interest.

Because few projects have well-defined guidelines for use in such decisions; because decisions like this are an every-day occurrence; and because the correct decision is usually impossible to ascertain by judgement, most project decisions are made without any way of knowing that they are correct. This, of course, can jeopardize the plant's profitability. How, then, can project decision making be improved? One practical way is to issue guidelines for project decision making which specify the relative importance of:

Investment cost
Operating cost (including fuel consumption and maintenance)
Performance characteristics (e.g., service factor)
Startup date

These guidelines can apply to small projects generally or to a specific project. Examples of guidelines for decision making are:

It is worth $4,000 in increased investment for each day that the startup date is advanced
A 1% improvement in service factor is worth a 6% increase in investment cost
A 1% reduction in operating cost is worth 4% increase in investment cost

These guidelines can be easily derived from the economic analysis that was done to justify the project. Although the techniques of project-profitability analysis are outside the scope of this book, they generally involve calculating the projected cashflow for the project's operating life and discounting it back to present-day costs. It is therefore a straightforward exercise to determine the sensitivity of the projected profitability to variations in the parameters of investment cost, time of startup, and operating cost. There are some cautions that should be heeded by users of these guidelines. Although they are useful, the guidelines can occasionally be abused by using them to justify irresponsible decisions. This is a problem that can be avoided by judicious monitoring and checking of the project decision-making process.

Life-Cycle Costs

The development of guidelines for decision making also introduces an important concept: that of "life-cycle costs." Life-cycle costs focus on the net operating cost of an item over its entire life, and include investment costs, energy consumption, maintenance cost, and operating efficiency. So, for example, if we specify a pump for a facility which is to operate for 30 years, we would normally select the pump with the lowest investment cost, as long as it met the design requirements of gallons per minute, pressure differential, etc. However, when life-cycle costs are considered, it might be apparent that a pump with a much higher investment cost would have to be shut down less frequently for servicing, be replaced less frequently, and perform more reliably resulting in a much greater profitability for the facility as a whole. Although project-management methods are not yet to the point at which life-cycle costing is common enough to be a simple tool for small projects, the use of management guidelines, as described above, is a worthwhile first step towards incorporating this type of thinking in project personnel.

ESTABLISHING PROJECT PRIORITIES

In the typical situation in which there are many small projects in progress at one time, there is often a lot that can be accomplished by simply establishing a system of priorities. In the absence of such a system, project engineers could be competing for resources and management attention in an unproductive way. For example, a critical project could be kept waiting while work is completed on a nonessential project. As in the case of guidelines for decision making, it is not always possible for the project engineer to assess the relative contribution of various projects to profitability, be they his own projects or those of others. This requires a management perspective, and it is therefore management that must either set priorities on projects or establish a method whereby a priority can be established on each new project. Some of the questions one should ask in setting priorities include:

To what extent is the project essential? (i.e., one should distinguish essential projects from merely desirable projects, and establish levels of importance)

Does the project involve an essential system in the facility? (i.e., systems, such as electrical, firefighting, instrumentation, etc., should be ranked in terms of how essential they are to plant operation)

What is the financial attractiveness of the project? (i.e., one should distinguish projects with a satisfactory return from those with outstanding returns)

Using these considerations, we might establish a priority system as follows:

Workscope priority[a]	System priority[a]	Financial priority[a]
1	1	1
2	2	2

Workscope priority[a]	System priority[a]	Financial priority[a]
3	3	3
4	4	4
5 (lowest)	5	5

[a]1 = Highest.

The Net Project Priority is the weighted average of priorities in each category.

Along with the system of priorities, there should be procedures for implementing the system. This is desirable to avoid the situation in which only high-priority projects are given attention, and the lower-priority projects are ignored until their slippage becomes so great that their priorities have to be upgraded. This, of course, is an abuse of the system which should be eliminated so that projects of every priority can progress together, and one should need guidance only when conflicts in the allocation of resources arise.

PERSONNEL CONSIDERATIONS

Some of the problems of small projects stem from the perception that project work is less important than operations, and this should be addressed by changes in the way the project engineer functions within the organization.

Continuity in the Project-Engineering Function

Some companies face the situation that the project-engineering function is used as a training ground for engineers who will, after a year or so, be transferred to other departments. Since most of the projects last less than a year, it sometimes seems that nothing is lost by this lack of continuity. A great deal is lost, in fact, and often the training that occurs is of questionable caliber.

Since small projects are in process continuously, some continuity in the project-engineering function is required to assure that something is learned from past projects and mistakes not repeated. Methods for planning, estimating and control, data, procedures, and systems should all be developed and maintained by the project-engineering function. If the project-engineering function had a core of experienced people whose main area of expertise was project work, and if proper methods, systems, and procedures were used in project work, then a year or two in project work would indeed be a valuable training experience.

Continuity in the project function also has the important benefit of fostering a sense of accountability. If projects are handled primarily by people who expect to be somewhere else in the near future, there is apt to be the feeling that mistakes will not become apparent until after the project engineer has been transferred. The old refrain, "it can't be my fault, everything was fine when I left" only serves to diminish the sense of responsibility and can significantly weaken the entire project-management effort.

Authority and Responsibility

Another frequent concern of the project engineer is that, though he is responsible for the successful outcome of the project, he has little authority with which to discharge that responsibility. Since he has to rely on other parts of the organization to provide the resources he needs (such an engineering services, drafting, purchasing, warehouse materials, and construction) he often has to deal with those who are higher in the organization than he is. This often makes it difficult to secure the necessary commitments to perform, and the necessary information for project control. While it would, of course, be inappropriate to make every project engineer a manager, it is possible to define clearly the project engineer's duties and responsibilities, along with the authority given to enable him to discharge those responsibilities effectively.

Project Engineer Performance Appraisal

Problems in Performance Appraisal

Appraisal of a project engineer's performance is often difficult. Theoretically, one should judge performance in terms of whether most projects are completed on time and within budget. In the small-project environment, this is not very easy, as small-project results can vary for the many reasons we have discussed, such as scope changes, lack of access, interferences, and unforeseen construction difficulties, which are not the project engineer's fault. And, if too much negative emphasis is put on schedule delays and cost overruns, the result is likely to be "fat" estimates and leisurely schedules, which, of course, do not enhance profitability in any way. What then can we do to establish performance creteria for the project engineer? How can we measure his performance against these criteria?

Use of Standard Methods and Well-Defined Objectives

If the methods by which projects are to be managed are reasonably well defined, and management's objectives are clearly expressed in a system of priorities and guidelines, then the consistent behavior that sets the standard has been defined. The degree of skill, judgement, maturity, and creativity that a project engineer displays in the conduct of his work can then become the criteria for performance appraisal. The methods described throughout this book will form a satisfactory basis for such a system.

By defining the system, and measuring performance in terms of how well the system is used, we avoid the unfair judgements that often occur. Companies that punish project engineers who preside over an overrun, and managements that make it a career-threatening move to go back for additional funds without regard to the reasons why (and such situations are very common) only foster a climate of dishonesty. Many project engineers feel like the Greek messenger whose reward for telling bad news was to have his throat cut. That is an experience one does not wish to repeat, even if it is in management's best interest to hear the bad news.

If a company has a policy that states that estimates should have an equal chance of overrunning or underrunning (which it should), it is to be expected that some estimates will overrun by more than 10% and therefore need additional funds. It is a serious occurrence when a project overruns, but the questions should be:

Was the project engineer effective in controlling the job?
Was the original budget plan reasonable, based on the information documented at that time?
Was the potential overrun revealed as soon as it became apparent, so that management action is still possible?
Are the reasons for the overrun things that could have been controlled, avoided, or compensated for by the project engineer?

Given a basis for a fair appraisal of this performance, most project engineers will not hesitate to report potential delays or overruns and, as we shall see in Section III, the willingness to make realistic forecases (even if unpleasant) is one of the keys to effective project control.

Management Shares in the Responsibility

Successful projects are the result of effective cooperation between project personnel and company management. Management's role in a project is:

To establish the methods, systems, and procedures for project management
To establish and administer the organizational structure for projects

Approve the budget plan for each project
To review progress reports and take action when appropriate

The first two items describe management's role in creating the means
to handle projects generally. These are described in "Establishing
Communication" and "Developing Data and Methods for Small-Project
Management," later in this chapter. The last two items describe management's role on each specific project, and are discussed below.

Approving the Budget Plan

The preceeding chapters have described the steps required to develop, present, and obtain approval of the budget plan. The budget
plan, or project model, consists of:

The *network plan* (what is to be done)
The *schedule* (when it is to be done)
The *resource plan* (what it will take to do it)
The *cost estimate* (how much it will cost)
The *documentation* (of the design basis and important assumptions)

Management's approval of such budget plan should mean more than
"OK, go ahead." It should also imply agreement that the budget plan
is consistent with the corporate objectives that the project is designed to meet, and that it has been prepared in accordance with established methods. Management who, after all, have the ultimate responsibility, should be seen as senior partners in the project-
management function.

Taking Action When Appropriate

There are apt to be conflicts, problems, and questions as the project
progresses which cannot be resolved at the project engineer's level.
These should be referred to management and resolved promptly. Many
projects are delayed due to the project engineer's not being able to get
an answer when one is needed.

ESTABLISHING COMMUNICATION

Small projects, like most other corporate activities, benefit greatly
from improved communication, particularly since so many diverse functions within the organization are involved. Some of the methods which
have been successful in improving project-related communication
include the following:

Project-Coordination Function

The project-coordination function consists of one or more people who serve as a coordination and communication channel between the project engineer and the other functional groups. The coordination function helps relieve the project engineer of that time-consuming part of his job. In the multiproject environment, the coordination function also helps to balance requirements between projects, thereby avoiding problems of resource allocation before they start. For example, if one project has slipped, the coordinator, who has an overview of all the projects, can easily see which project could best use the resources which have become available.

Small-Project Group

Some companies have found it worthwhile to set up a "small-project group." The purpose of this group is to integrate all the organization's functions related to small projects and, where possible, perform those functions itself. For example, the small-project group might have a small-project engineering function, so that a certain amount of straightforward design and drafting work could be done without recourse to the design and drafting department. Similarly, some purchasing and subcontracting functions could be placed within the group.

Status Review Meetings

Meetings to review the status of the small projects, resolve problems, allocate work assignments and resources for the coming weeks, and advice management have proven to be worthwhile. If the project-reporting systems have been well defined, these can be used as the basis for discussion so that each project's status can be reviewed quickly.

Formal Reporting Procedures

In order to assure that management is kept informed, formal reporting procedures must be implemented. With the advent of word processors and microcomputers this is much easier to accomplish than it was even a few years ago. Chapter 7 discusses management reporting in some detail, but the point to be made here is that the importance of sound and timely project reporting needs to be recognized by management, and the reporting effort supported by attention and prompt action when needed.

Encouraging an Enlightened Project Management Style

Most texts on management describe two general management styles:

Theory X

A management approach which assumes that those who are being managed dislike work, will seek to minimize the amount of work they do, have little ability to manage their own work, and little ambition.

In the case of Theory X, organizational goals are in conflict with individual goals, and sufficient effort to achieve organizational objectives can only be obtained by tight control and the threat of punishment.

Theory Y

A management approach which assumes that those who are being managed like to work, will work hard and exercise self-direction in order to achieve job satisfaction, will naturally seek responsibility, and will utilize ingenuity and creativity in solving problems.

In the case of Theory Y, organizational goals are more closely aligned with individual goals.

Most project management has a strong bias towards Theory X, especially where client vs. contractor relationships are concerned. In order to move to a working relationship based more on Theory Y, there have to be common goals as well as measurements of performance. Thus, the methods and systems for planning, estimating, and control which are discussed in this book can be thought of as a basis for better communication, and hence a means of gaining a project-management style more like that of Theory Y.

DEVELOPING DATA AND METHODS FOR SMALL PROJECT MANAGEMENT

If effective project-management practices are to be implemented in an organization, the same integrated approach that works so well on projects should also be considered for methods development. If, as is so often the case, an organization tries to improve its performance in one area of project work, that improvement is apt to be diminished by a lack of improvement in other areas. So improvements are best addressed in a comprehensive manner which should include the following basic elements:

Databases for collecting, storing, and presenting data
Information coding systems
Planning and scheduling methods

Cost estimating methods
Guidelines for contracting
Quality assurance programs
Project control procedures

Each of these elements is described below.

Developing a Project Database

Why a Database for Small Projects?

A database can be thought of, generally as a file of information organized in a useful way that provides the capability to present the
data according to the user's requirements. A database provides specific, discrete information, as distinct from a method which provides
generalized relationships and/or ways of operating. The data base can
be in the form of a reference book, a set of files, or can be computerized.

Since almost all the data used in planning and estimating projects is
empirical, the use of a database can contribute a great deal of consistency, accuracy, and credibility. Since most small projects occur
in significant numbers, collection of data is relatively easy. So, setting
up and maintaining a database for planning and estimating is a step
which will greatly improve the small-project management function at
minimal time and cost.

A database for project information also helps to assure that valuable
project data is captured and made available for the next project. A
classic problem with project data is that there never seems to be enough
time to prepare final reports and see to it that the learning from the
project at hand is documented and passed on. Too often, the pressures of the next project are too great, and the company is thus des-
tined to repeat it's mistakes. Fortunately, if the database is computerized, a lot of the data can be captured automatically.

Designing the Database

The most difficult aspect of a database is its design. To design a
database, we must be able to specify the following:

The data required for planning, estimating, and resourcing. For example, what type of estimating do we wish to do? How much detail is
required? Will we need piping costs for each flange and fitting, or
will cost per ton of piping be sufficient?

The format that the data is generally available in. For example, do
we generally receive field reports showing man-hours or do they show
the number of people?

Other uses to which the data will be put. For example, can we use man-hour data which is already available in the accounting department?

The independent and dependent variables. For example, cost is usually dependent on independent-design variables such as pump horsepower, tons of steel, etc. How should they be related in the database?

The number and relationship of variables that are to be handled at once.

The facilities that will handle the database. Will it be manual or computerized? What limitations exist as to volume of data handled?

Some typical databases are shown in Figure 6.1.

The design of a computerized database must also specify how data is to be processed, in terms of input (via keyboard, cards, tape, or disc), internal processing, and output (visual displays, printed reports, tape or disc storage). There are three distinct types of structure for a computer database:

Heirarchal database: This locates information in progressive levels of detail. For example, a post office zip code locates the area of an address by progressively defining state, city, area, and group of streets.

Relational database: This locates information by relating "key fields" in one database to another.

Network database: This allows a "many-to-many" relationship of data fields.

As we extract information from various files ("datasets") in the heirarchal database, the system proceeds down through the heirarchy each time. A relational or network database allows us to extract data from each file without using the heirarchy. When evaluating computerized databases, the ability to select, sort, aggregate, arrange, and display data the way we want to is the most important feature, and the database structure helps determine how much flexibility we will have.

Setting up the Database

Although a database can be manual or computerized, there is currently little reason to set up and maintain a database by hand. There are many good programs for data management, including those for personal computers. For small projects, these simple and inexpensive databases work fine. Often, the existing in-house computer can be used. This is particularly appropriate when input data is already available in the system.

When setting up the database it is essential to have an information coding system to organize and recall the data. This is discussed in

PLANNING

WORK CATEGORY	ACTIVITY	I QUANTITY	I MANHRS./UNIT	I CREW SIZE	I REQ'D EQUIP.
FOUNDATIONS	EXCAVATE	I C.Y.	I	I	I
	FORM	I C.Y.	I	I	I
	REBAR	I C.Y.	I	I	I
	POUR	I C.Y.	I	I	I
	CURE	I C.Y.	I	I	I
	BACKFILL	I C.Y.	I	I	I
		I	I	I	I
PIPEFITTING	UNLOAD	I SPOOL	I	I	I
	SHAKEOUT	I SPOOL	I	I	I
	ERECT & CLAMP	I SPOOL	I	I	I
	WELD	I DIA.-IN.	I	I	I
	TEST	I WELD	I	I	I
	REPAIR	I DIA.-IN.	I	I	I
	LINE FLUSHING	I LINE	I	I	I
		I	I	I	I

(a)

ESTIMATING

COST GROUP	CATEGORY		I QUANTITY	I COST $/UNIT I
PIPING (CS)	PIPE	10 IN.	I LIN. FT. I	I
	PIPE	12 IN.	I LIN. FT. I	I
	PIPE	16 IN.	I LIN. FT. I	I
			I I	I
	ELBOWS	10 IN.	I EACH I	I
	ELBOWS	12 IN.	I EACH I	I
	ELBOWS	16 IN.	I EACH I	I
			I I	I
	TEES	10 IN.	I EACH I	I
	TEES	12 IN.	I EACH I	I
	TEES	16 IN.	I EACH I	I
			I I	I
CENTRIFUGAL		5	I HP I	I
PUMPS		10	I HP I	I
		25	I HP I	I
		50	I HP I	I
		100	I HP I	I
		500	I HP I	I

(b)

FIGURE 6.1 Typical small-project databases. (a) Planning. (b) Estimating.

the next section. The database must be on a consistent basis throughout, and that basis must be clearly defined in terms of the conditions
assumed, performance by client and contractor, pricing and productivity levels, "base date" for costs, etc.

A database design often consists of several databases which can be
used separately or together. For example, if the integrated approach
is to be used for planning, resourcing, and estimating, the databases
of man-hours, durations, and costs should all be related to assure that
consistent results are produced.

An important function of a database is to collect raw data from projects and to use this data to assure that the database is up-to-date. If,
however, raw data was fed directly into the database, there would be a lot
of confusion as a result. Raw data must first be analyzed to put it on the
same time-and-cost basis as the rest of the data in the database. This
process of normalization requires judgement, and a knowledge of the
circumstances surrounding each project. A simple data system design
is shown in Figure 6.2, illustrating the *holding database* in which raw
data is stored, as well as the *reference database*, which is used for
planning and estimating. The reference database should be protected
from unauthorized modification, and should be updated every 3 to 6
months, based on analysis of data from the holding database.

Sources of Data

The quality of the data is, of course, the main consideration of a database and, in the initial stages, can be a problem. Bearing in mind that
the best data is that which best predicts what will happen on a new
project, it can be seen that historical data is not always the most desirable. For example, we may be aware that the last 4 turnarounds
were on schedule, but used excessive man-hours and overran their
budgets. Should we use data from these projects in planning and estimating the next one? To do so would imply that the unsatisfactory
past performance is something we expect to repeat. Not to do so could
be interpreted as a form of "cockeyed optimism," which has no place
in budgets.

Since our purpose is to predict that outcome which has a 50/50
chance of being achieved, we should predict, with our plan and estimate, that the current performance will be better than the past if that
is what we expect to happen. To make that judgement, we should know
why the past projects encountered problems, and why the current one
can be expected to avoid these problems. Having determined those
answers, we can make adjustments to the data and thereby establish
the basis for future estimates. However, if analysis of the past data
shows that we cannot reasonably expect the performance to be improved, then the unaltered past data is the best predictor of future
performance.

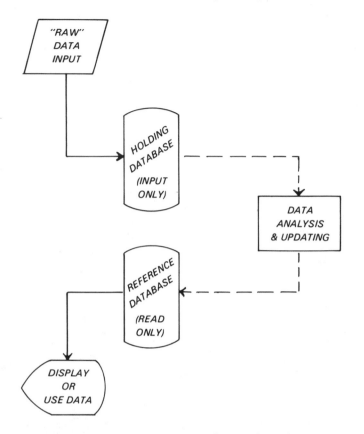

FIGURE 6.2 Use of databases for project data.

The above illustrates the amount of work that goes into defining and loading data. An additional source of work is obtaining the data in the first place. Unless past projects have been managed with an eye towards collection of data (and how many are?) we are likely to find that the collection, organization, analysis, and normalization of data can be a daunting proposition. It is important that it be done right, since all future projects will be based on the data we put in. Although the preferred method is to use actual past data, even if it is adjusted, there are other ways to load the database if the use of actual data turns turns out to be impractical. Many companies have solved this problem by accessing a database which is available commercially. Some of these databases are quite good, and there are a number available for architectural applications as well as process-industry projects. The problem is that the industry average may not fit the typical small project in a given company. Although this can be easily addressed through the use of correction factors, it takes some time to establish those factors by comparing actual vs. predicted results.

One approach that works well is the use of "synthetic data." This involves creating data which represents what is expected, even if it does not correspond to actual past data. At worst, synthetic data will give results that are about what would have been estimated anyway. As actual data is collected, normalized, and used to update the database, the synthetic data is replaced by actual data. To develop synthetic data, standard or typical designs are produced for a variety of design conditions. The expected cost is estimated for each case. Similarly, standard networks are prepared and resourced for each design case. The data is then broken down into the detail required to fit the database format.

Designing the Information Coding System

What Is It?

The information coding system organizes the flow of project data. Although it might, at first, appear to be a simple task, the design of a coding system for project information is quite difficult, particularly for small projects in an operating plant, because so many different organization functions are involved. The functions that deal with project data include:

Contract administration
Purchasing
Planning and scheduling
Cost engineering
Project engineering
Accounting
Field supervision
Documentation
Quality assurance
Company management

On small projects, some persons may perform more than one function. The coding system gives each person and each function the information he needs, when he needs it, in the format with which he is comfortable.

Coding System Requirements

The project-information coding systems should be:

Simple to use (i.e., long, complex codes are to be avoided)
Compatible with the company's codes and those used by contractors

Flexible enough to sort, select, and present information according to
user requirements

Compatible with the constraints imposed by the computer system to
be used

Capable of handling an integrated approach to planning, estimating,
and resourcing

To meet these requirements, coding systems often function in a
heirarchy of codes and subcodes, such that excess detail is not a problem to each user.

Elements of the Information Coding System

The basic elements on which the coding system is based are:

The network plan
The cost coding system
The organization chart
The work breakdown structure (if applicable)
The design coding system
The accounting system

The coding system manages information according to the specifications of these basic elements. It allows us to manipulate the project
model, and prepare reports using cost data, resource data, or scheduling data. It should work as follows:

The network plan should have a code for identifying activities for
use in scheduling, resourcing, and progress measurement

The resources should have a cost code, identifying them by labor
craft, material type, etc.

The organization chart specifies who gets what information and in
what format

The work-breakdown structure (WBS) relates the network activities
to specific contracts or work packages. This technique is very
useful for large projects and is mentioned here for those who are
already using it. It is probably not appropriate to implement WBS
for small-project use so it is not explained further in this book.

The design code should be used to indicate what system an item of
material belongs to, or what type of material it is.

The accounting code must be compatible with the cost data received
from the project.

By examining each of these elements, and combining them wherever
possible, a practical coding system can be developed. A simple but
effective coding system concept is illustrated in Figure 6.3. The

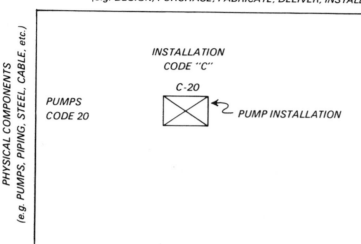

FIGURE 6.3 Basis elements of a project-information coding system.

coding system is seen here is a matrix in which the rows are physical components of the project, and the columns are the activities (or tasks) that are performed on those components. The cell, then, describes a "cost center" which relates the physical component to the tasks of the project. Should additional detail be required, the matrix can be extended into additional dimensions.

Developing a Method for Planning and Scheduling

The technology of planning and scheduling was discussed in Chapter 3. To implement such techniques in-house, an organization has to first define its needs and constraints so it can develop specifications for its method. Some of the points which should be considered are:

The specific organizational needs which the method must satisfy
The specific methods to be used (e.g., ADM or PDM, CPM or PERT)
The formats and nomenclature to be used (i.e., the definitions of levels of networks and when each should be used)

The data required (we must assure that the database will provide it)
The use of standard networks for similar projects
The procedures for preparation and review
The amount of training required
The computer systems which are available or will be required
The interfaces between the planning and scheduling methods and
    those for resource planning, cost estimating, and project control.

Having defined the specification for the method, it can be developed
and implemented in the following steps:

1. Establish database and coding system (see sections above)
2. Prepare standard or sample networks for training
3. Define computer requirements, and select and implement system
    (see Chapter 11)
4. Prepare procedures
5. Select the first project to be planned with the new method
6. Plan the first project using the new method in parallel with
    existing methods
7. Evaluate the results, and modify as required
8. Obtain management approval for implementation
9. Train users

Developing a Method for Cost Estimating

As in planning methods, the development of a method for cost estimating begins with a definition of company needs so that one can define
a specification for the method. The steps that should be taken are
shown below.

Definition of Estimate Stages

Typical stages of a project during which estimates are prepared are:

Screening, during which projects are compared and some selected to
    progress further
Design optimization, during which the cost of alternative designs
    are studied
Project approval, during which the approved budget is established

It is evident from the above that, in the case of the screening and
optimization estimates, relative costs are more important than absolute
costs. The estimate prepared for project approval, however, must be
as accurate as possible. Small projects in some companies are in
simpler situation of preparing only one estimate: the budget estimate.

Definition of the Estimating Basis at
Each Stage

In order to know what estimating method will be most appropriate at
each stage, the degree of definition of the estimate basis should be
defined at each stage. Consideration should be given to:

   Design: To what extent has the design progressed at each stage?
      How much of the design basis is apt to be preliminary? How much
      will be firm?
   Planning: How well defined are the plan and schedule likely to be?
   Cost data: What type of data is likely to be available?

The estimating method must be appropriate to the basis definition as
well as to the use to which the estimate will be put.

Definition of Estimate Use at Each Stage

Estimates help in decision making in a number of different ways, including:

   Deciding which projects are worthy of further work, developing pre-
      liminary budgets, and performing preliminary economic analysis
      (at the screening stage)
   Deciding which design or planning approaches are most desirable,
      refining budgets, and obtaining approval to proceed with further
      design work (at the optimization stage)
   Obtaining project approval, budgeting, and setting up cost control
      procedures (at the approval stage)

Definition of the Required Accuracy at
Each Stage

Based on the use to which the estimate will be put, we can define the
accuracy required for each type of estimate. The accuracy which can
be achieved is, of course, only as great as that of the basis. It is important to recognize that estimate accuracy cannot exceed the accuracy
of the basis, and therefore reasonable expectations of estimate accuracy
are in order.

Development of Design-Cost Correlations

The cost-estimating method is intended to provide generalized relationships between cost (the dependent variable) and independent variables
such as design quantities. Once the database is established, the method, either in the form of estimating curves and tables or computerized
equations, can be established through the development of cost correlations. The design of the method, i.e., what correlations are

established and their statistical accuracy, should match the specifications developed in the above steps.

An important point in methods development is that a certain amount
of inaccuracy is acceptable. Many estimators only feel comfortable with
a lot of detail and it is a common mistake to prepare estimates in ex-
crutiating detail which contributes little to overall accuracy. This can
be easily seen in the case of estimates which are escalated over significant time periods: the escalation factors are bound to be relatively
inaccurate, yet the escalation cost may be as large as the base estimate
itself. So the extra effort required to add accuracy through increased
detail (which increases estimating time and cost) should be evaluated
carefully to see whether the added accuracy is significant and
necessary.

Documentation and Training

As in the planning method, final steps in implementing a planning
method include documentation, preparation of procedures, management
approval, training, and testing on a sample project.

Guidelines for Contracting

The various approaches to contracting are described in Chapter 3.
It becomes evident there that there is a good deal of judgement involved in selecting the right approach for a particular project and for
the company in general. Because contracting decisions can involve
considerations, such as legal aspects, that are outside the realm of
most project engineers, it is important that management guidelines be
provided. These guidelines should indicate what type of contract is
preferred for each type of situation, and what factors should be used
to decide on a contracting approach.

Once the project engineer has selected the type of contract to be
used, the guidelines can also provide typical contract forms which have
been reviewed by the legal staff and approved for project use. This
avoids any dangerous legal implications from the blunders of inexper-
ienced project engineers, and streamlines the approval process.

Procedures for Quality Assurance

A complete set of project procedures must include quality assurance.
As discussed in Chapter 3, the quality assurance program must be
well planned. Like contracting, decisions regarding design quality
(see "Guidelines for Decision Making," earlier in this chapter) often
require an appreciation of management considerations that are not generally available to the project engineer.

The quality-assurance procedure should be the result of an in-depth
design and cost analysis, based on life-cycle costing as discussed in
"Life-Cycle Costs," earlier in this chapter. The first priority is to

define quality in specific design terms, such as those used in reliability engineering (e.g., "mean time between overhauls," and "mean time to replace"), as well as those relating to operating efficiency and maintainability. A lot of work has been done in recent years to bring this type of engineering, developed in the aerospace industry, to industry in general. Once these specific design requirements are defined, they can be included in the bid package for equipment thereby assuring that the desired quality is obtained.

Quality assurance also requires an effective change-order procedure, since many design improvements are brought into the job as changes (see Chapter 10). If changes are properly evaluated, according to guidelines as discussed in "Guidelines for Decision Making," the optimum blend of design quality and investment cost can be achieved.

It should be remembered, when discussing quality and cost, that the design of a project is the aspect over which we have the most control and that design has the most profound effect on cost. Therefore, any procedure which helps assure that the cost implications of design decisions are considered will go a long way to assuring a profitable project.

Project-Control Procedures

The principles of project control are discussed in detail in Section III. It is, however, appropriate to include project control in this discussion, as it is the means by which management objectives are achieved. In fact, everything that has been discussed in the first 6 chapters of this book provides a proper basis for project control, i.e., organizations, methods, data, systems, and procedures for planning, resourcing, estimating, quality assurance, and contracting. We now have something against which we can measure: a comprehensive, integrated plan.

If company management has provided the tools for planning and budgeting the project, it will also want to make sure that the systems and procedures are in place for controlling it so that the desired results are achieved. The steps that are required to develop and implement project-control procedures include:

The definition of the objectives (i.e., what each procedure is to accomplish)
The description of the method (i.e., what basic techniques are to be used)
The allocation of responsibility (i.e., who does what to whom)
The definition of the flow of data and information
The preparation of an estimate of time and cost to implement
The gaining of management approval
The development and implementation of computer systems (if needed)
Documentation

Testing on a sample project in parallel with existing procedures
Modification as required
Training
The briefing of contractors and revisions to contract documents
Complete implementation

CHAPTER SUMMARY

This chapter covered management's responsibility in the successful
completion of small projects. Much can be done internally to assure
that the necessary data, methods, systems, and procedures are available to the project engineer. Some simple modifications to the organization can help a great deal as well.

With company management and project engineering now armed with
a comprehensive, integrated plan, a schedule and estimate, and an
approved budget, we are ready to discuss the part most of us enjoy
the most: the execution of the work. The next four chapters, which
comprise Section III, explore the various aspects of project control,
and describe techniques which will help assure that the final results
produce the profitability that is expected.

Part III
Project Management and Control

7

Management Reporting: The Key to Project Control

PROJECT CONTROL: A MATTER OF PERSPECTIVE

Many project engineers look at a project and see the large number of people performing various tasks, discussing problems, holding meetings, writing memos and reports, making telephone calls, etc. In the field, of course, they see lots of construction equipment, more people, and more problems. Their day is often spent in going through the ever-overflowing in-tray, returning phone calls, attending endless meetings, plowing through voluminous reports, preparing voluminous reports, and just generally trying to keep up with what is going on. Contrary to popular belief, very little control over a project can take place in this environment.

Everyone who has flown in an airplane has noticed how different things look from that lofty perspective. The snarled traffic which we fought through to get to the airport now seems a more curiosity if we notice it at all. We see our cities and towns in a much broader context from the perspective of the air. Successful project management also requires a different perspective: The ability to rise above the day-to-day rush of activities and get a clear picture of what is really going on. So, how do we get the "airplane" perspective on our projects?

Effective project control requires only that we know, at all times, the answer to four very simple questions:

If current trends continue, when will the project be complete?
If current trends continue, what will the project actually cost?
What are the causes of adverse trends?
What is being done to correct adverse trends?

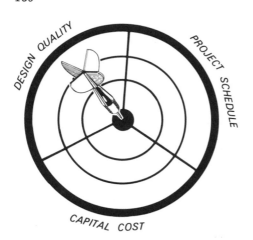

FIGURE 7.1 Project-management target. This shows that the optimum combination of quality, schedule, and cost is that which produces the maximum profitability.

These questions can be answered by the collection, analysis, and presentation of project data. Data, properly presented, can give us the airplane view we need to raise our sights from the detailed day-to-day activities to the target of those activities: the optimum blend of cost, time, and quality. Project data gives us the feedback we need to adjust our sights; to define, aim at, and hit the "target" which is the optimum combination of project variables (see Figure 7.1). If our current direction is off-target, we should be able to see why, and what can be done to correct our aim. The way in which data can be used for this purpose is discussed in this chapter.

EFFECTIVE CONTROL REQUIRES EFFECTIVE REPORTING OF PROJECT DATA

Project control takes place when those who are in a management or supervisory role are aware of the need for corrective actions, take the necessary actions, and followup to assure those actions were effective. The individuals who are in a position to effect project control include project managers, project engineers, design supervisors, purchasing agents, construction superintendents, craft supervisors, foremen, and anyone else who has the authority to affect the way people are working, either inside or outside the company. Even the company president can occasionally become involved in a project-control activity, particularly in cases in which resolution of problems with a contractor or supplier must take place at a high level.

The actions these individuals take are generally of a managerial nature, that is, they affect the way other individuals and organizations do their work. These actions may take the form of memos and meetings to discuss specific problems and develop solutions, as well as more drastic steps such as major changes to the contracting or construction plan. But what prompts a project-control action in the first place? The answer is *information*, namely:

What was expected to happen? (the budget plan)
What did happen? (data representing work to date)
How has actual performance varied from the budget plan?
What trends can be identified from performance to date?
What are the forecasted final cost and completion date, based on performance to date?
What problems can be identified from analysis of performance to date?
What are the possible corrective actions that can avoid delays and overruns?

So it is evident that the quality, quantity, and timeliness of management information is the key to effective project control. We can summarize this concept by saying that project control is simply a matter of presenting:

The right information
In the right format
To the right person
At the right time

This concept is illustrated in Figure 7.2, which is described as follows:

1. A project actually consists of many people doing a lot of different things, with various results. This is the "real-life" situation which we represent with data. For example, progress is measured by data summarizing the total work accomplished and man-hours representing the hours actually spent; the relationship of man-hours spent to work accomplished indicates productivity. So the first step in project control is capturing data which will represent what is really going on.

2. Once we have our data, we subject it to analysis, that is, we arrange it in a certain format and compare it to our original plan to identify variations to date and/or anticipated variations. For example, we might summarize our man-hour expenditure data for comparison with

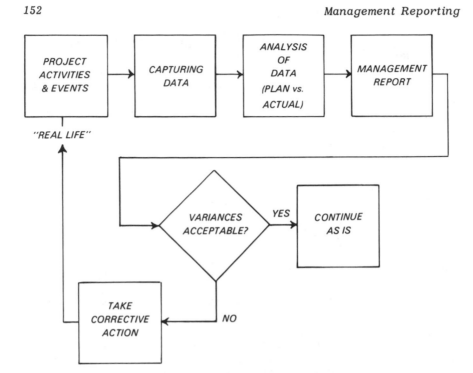

FIGURE 7.2 Elements of project control.

the budget plan, and calculate the variation in the actual man-hours
expended to date for each labor discipline, against the planned man-
hour expenditure.

3. The analysis of data gives us an indication of how the project
is progressing. This assessment of the project's status and outlook
is summarized in a report. Preparation of this management report
usually involves a narrative description of project status, tables of
data to pinpoint the facts, and graphical displays to summarize the
data. For example, we might note that the man-hours expended in
electrical work were greater than planned, while the physical progress
is less than planned. Our narrative would call attention to this fact,
point out the cost and schedule impact if nothing is done, and suggest
corrective actions that could be taken.

4. After reading the reports, management has a decision to make:
are the status and outlook as presented in the report satisfactory?
If so, nothing special needs to be done and the project can continue
as it was. If, however, the results to date and/or the anticipated
final results are less than management expects, some corrective action

is required. The report should, in fact, highlight the problem areas and recommend corrective action. For example, corrective action to offset the adverse trends in the electrical discipline might be to review the status of delivery of electrical materials and drawings to the field, the performance of the electrical subcontractor, and the coordination of his work with that of the other crafts.

5. The corrective action is then followed up to assure that the desired impact on the real-life situation is achieved.

So we can see that management reporting, far from being a dull, after-the-fact recitation of figures, is the key to effective control.

TIME: THE MOST CRITICAL ELEMENT

For management reporting to be an effective means of project control, it is essential that the time to complete the "loop" of information flow (as shown in Figure 7.2) be as short as possible. On large projects, it often takes four to eight weeks to progress through the loop, which implies a period of up to 12 weeks from the time something happens to the time something may be done about it. For small projects, which may be completed in only a few weeks, this is clearly unacceptable. In fact, for small projects, speed of reporting is more important than accuracy since, if the information is not available when timely, it will be virtually useless. This means that control of small projects requires a very short cycle for management reporting. To accomplish this short cycle, we will use a combination of shortcut techniques as well as computer-assisted methods.

THE BUDGET PLAN PROVIDES THE BASIS
FOR CONTROL

The budget plan (developed in Chapters 3, 4, and 5) serves as our "road map" for the project-execution phase. In other words, it describes every aspect of how we intend to utilize resources and time to accomplish the given scope of work.

When using a road map to navigate, we compare the map's information on what we should see (towns, roads, etc.) with what we actually see as we drive along. We use this information to calculate where we are, on the map, and compare that to where we planned to be. In this way, we are able to deduce whether or not we need to correct our course, and, if so, how to proceed. We can also predict what will happen if we do not change our course, as well as the time we expect to arrive at our destination, and whether we will be late. The budget plan is used in exactly the same way. As the project progresses, we monitor our progress relative to the budget plan, comparing "where we are" in terms of physical progress, time, man-hours, and cost, with where we expected to be. When we seem to be "off course," we can use our budget plan as a guide as to how to proceed. The difference between the planned and actual values of the project-control parameters is the "variance," and it is through the identification and analysis of variances that adverse trends and problems are identified, and control effected.

The project model, representing the design, planning, scheduling, resourcing, and cost basis of the project, can be used to identify variances and to forecast the result if they are allowed to continue. The budget plan is our fixed reference point, and can be referred to as a "static model." The project as it stands today can be represented by a "current model," reflecting progress, man-hours, productivity, and changes to date. Our current projection as to the final outcome of the project can be represented by a "forecast model." If these integrated models are computerized, it is relatively simple and quick to update the data and prepare useful reports in which the three project models are compared.

DEFINITION OF PROJECT–
CONTROL TERMINOLOGY

Before proceeding further with the techniques of project control, it is important to define our terms. Like most other aspects of project management and cost engineering, the terminology is often ill defined, and used inconsistently. The definitions shown below are those that the author has found to work well in practice.

Physical progress: A measure of the amount of work done to date, based solely on physical quantities, expressed as a percentage of the current total scope of work.

Man-hours spent: The total direct man-hours which have been spent.

Productivity: The ratio of planned man-hours to accomplish a given scope of work to the actual man-hours spent.

Learning curve: The measurable tendency of the time required to perform a given task to decrease as the number of times the task is performed increases, until a maximum efficiency is reached.

Direct man-hours: Man-hours which contribute to measurable physical progress.

Indirect man-hours: Man-hours which are required to support the project activities, but which do not contribute to physical progress.

All-in hourly rate: A cost per man-hour which includes some overhead and/or indirect as well as direct costs.

Committed cost: The amount of money which should be set aside to cover the forecast final cost of all current purchase orders, contracts, and subcontracts associated with the project.

Cancellation cost: The amount of cost which would be incurred if the project were cancelled and a fair settlement made on all current contracts.

Expenditure: The amount of money that has actually been spent on the project, i.e., the total of all company outlays to cover project charges.

Value of work done: The total cost that would be incurred if all the work done to date were to be paid for according to contract.

Design change: A change, initiated by the design function, which alters the specific way the design is executed. Design changes are experienced on virtually all projects, and provisions should be made for them in the plan and estimate.

Scope change: A change to the basic specification of the project. A scope change adds facilities or capabilities that were not previously part of the project. Scope changes are generally not provided for in the plan and estimate. Note that a scope of change to a contractor, increasing his assigned scope of work, may not be a scope change to the project.

Field change: A change made in the field to facilitate construction. All projects experience field and startup changes, and these should be allowed for in the plan and budget.

Startup change: A change made in the field to facilitate the startup or operation of the facilities.

Punch-list: A list of small jobs that must be done before the unit is considered complete and ready for startup (also known as a "but-list").

Contingency rundown: A systematic reduction of the contingency included in the cost forecast, to reflect the addition of changes and the reduction of uncertainties.

Cost forecast: A prediction of the final cost of the project if present trends continue.

Trend curve: A curve which plots project parameters, such as progress and time, and, by using extrapolation and/or a standard curve shape (such as an S curve) can predict the final result.

"S" curve: A standard curve describing project parameters over the project's life. It is usually applied to large projects.

GUIDELINES FOR MANAGEMENT REPORTING
FOR SMALL PROJECTS

Use Management by Exception

On small projects, we can expect that any management attention we receive will be "by exception." In this context, "management by exception" refers to the fact that systematic, in-depth management review of the status of small projects is not likely. What is likely is that management attention will be obtained when their attention can be focused quickly on those items that need it. This is equally true for the project engineer, whose attention is often divided between many projects and the various day-to-day problems of each. Management by exception is made possible by reporting formats which highlight problem areas.

Management Reports Must Be Easy to Prepare

Due to the lack of both time and resources, management reports must be easy to prepare. That means that the reports must be free from any information which is nonessential or noninformative. The report should be complete, but just complete enough to advise of general status and convey the information needed when corrective action is required. To be easy to prepare, the report must utilize, wherever possible, input data which is easy to obtain and process.

Computerization of Report Preparation

In order to achieve the rapid processing of data and preparation of reports, computer-assisted techniques are very appropriate in the small-project environment. A small computer system can also produce significant improvements in report preparation time, through the use of word processing.

One feature on many computer systems, which can help in accomplishing management by exception, is the ability to sort, select, and order project information. For example, a program can be set up to produce a report which summarizes all the current projects and lists the projects in decreasing order of overrun amount.

Another way in which computer systems can help shorten the time required for report preparation is by maintaining a current file of relevant information, such that the report preparation requires little more than a printout of the current data.

Many of the current systems are designed to produce progress reports, and feature integrated software for word processing, "spreadsheet" calculations, data storage, management graphics, planning/ scheduling networks, and barcharts. More sophisticated systems can customize report formats, and tie the computer directly into other company systems. Computer applications are discussed fully in Chapter 11.

TRACKING CURVES: A USEFUL TOOL FOR SMALL PROJECTS

Tracking Curves: A Good Shortcut Approach

A tracking curve is the means by which we can show:

The variations between planned and actual performance
The forecasted final result, if nothing is done to correct current variations

Tracking curves have the additional advantages of being easy to prepare and graphical in their presentation, so that a great deal of useful information in conveyed quickly. They are, therefore, a very useful tool for small projects in which simplicity and time are of utmost importance.

What Is a Tracking Curve?

The purpose of a tracking curve is to represent, graphically, the variation in performance to date from the planned performance. By "tracking" performance to date, the curve also permits extrapolation to forecast the final results if present trends continue. Since microcomputer-based spreadsheet and graphics programs are widely used, and are highly appropriate for small projects, the illustrations in this (and other) chapters have been prepared using typical hardware and spreadsheet software in order to illustrate the various applications. The basic elements of a tracking curve are shown in Figure 7.3, and are described as follows.

The Dependent Variable

The dependent variable, drawn on the y-axis, is the project-control parameter whose performance and trends we wish to monitor, forecast, and control. Typical project-control parameters include:

Physical progress (i.e., % complete)
Man-hours expended
Cost of work done
Productivity
Unit costs
Hourly costs
Number and cost of changes

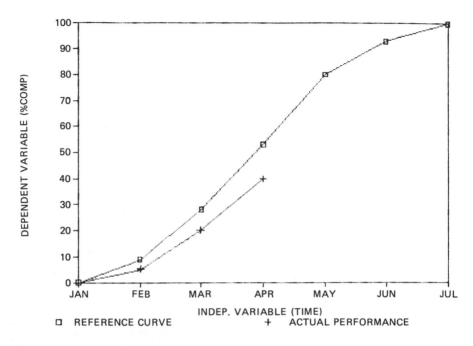

FIGURE 7.3 Elements of a tracking curve.

The Independent Variable

The independent variable, drawn on the x-axis, describes that project-control parameter which we wish to use as a measure of "where the project is." The two most common independent variables are:

> Time (i.e., days, weeks, months, or dates)
> Progress (i.e., physical % complete)

The Reference Curve

The reference curve, often referred to as the S curve, describes the relationship between the dependent and independent variables which has been anticipated by our budget plan. It might show:

> The rate at which we expect to make progress (i.e., the % complete per month)
> The rate at which we expect to expend man-hours or costs

The reference curve often takes the form of an "S" on large projects, because large projects often experience first a period of accelerated

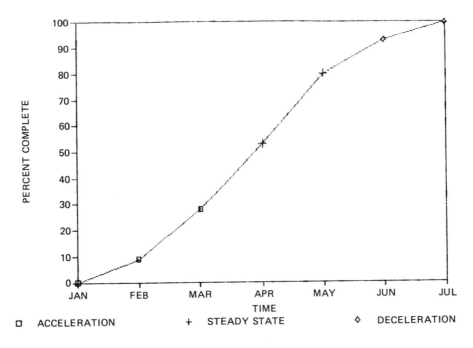

FIGURE 7.4 S curve for a large project.

progress, then a steady-state rate of progress, and finally a decelerated rate of progress. This is illustrated by Figure 7.4, which shows the stages of construction on a large project, and is described as follows.

1. In the *acceleration stage*, civil work predominates. Site clearance, excavation for foundations and underground lines, installation of temporary construction facilities, and construction of foundations all occur during this phase. The rate of progress is often limited, because the other craft operations (mechanical and electrical work, ironwork, etc.) cannot begin until the preliminary civilwork is almost complete, and there is a limit on just how quickly that work can be progressed. The manpower on the job usually builds up during this stage.

2. In the *steady-state stage*, all crafts are able to work on the job. Manpower reaches a peak during this stage, and remains relatively constant. The rate of progress also reaches a peak and remains relatively constant as well.

3. In the *deceleration stage*, the work of the major crafts is essentially complete, and manpower is reduced accordingly. The rate at

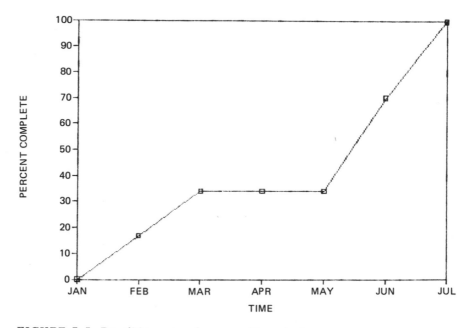

FIGURE 7.5 Possible curve for a small project.

which progress is made also, of course, decreases. The type of work done during this stage (painting, insulation, startup changes, "punch list" items, etc.) also does not appear to contribute much to measurable physical progress.

But what of the small project? What shape does its reference curve take? For a small project, it is important to recognize that *the reference curve will not necessarily be an S curve.* The small project may not require all the various crafts or have a period of manpower buildup or rundown. In fact, many small projects, such as turnarounds or other maintenance operations, have a constant number of people from start to finish (see Figure 7.5).

Tracking-Curve Format

There are several different formats available for tracking curves, the selection of which depends on the type of variables being used, the preference of the user, and any computer hardware or software limitations that might apply. The different configurations are illustrated in Figure 7.6, and are described as follows.

FIGURE 7.6 (a) Cumulative progress curve for a small project. (b) Incremental progress curve for a small project.

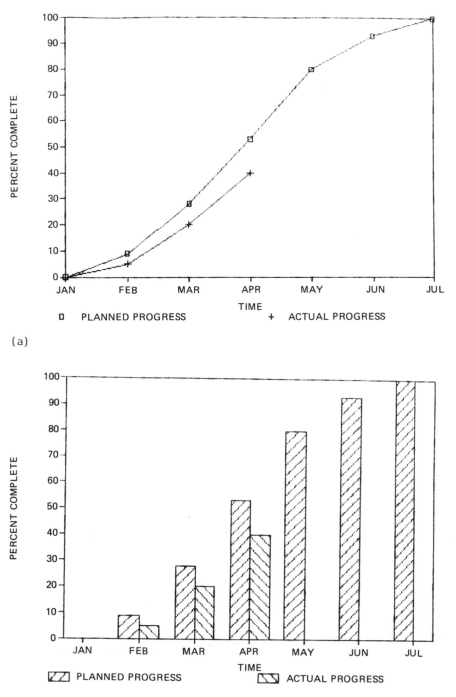

(a)

(b)

Cumulative

This illustrates the cumulative (i.e., up to-date) value of the dependent variable (see Figure 7.6A). For example, the dependent variable might be "percent complete earned to-date." This type of curve is usually continuous.

Incremental

This illustrates the incremental (i.e., for-the-period) value of the dependent variable (see Figure 7.6B). For example, the dependent variable might be "percent complete earned this week." This type of format is usually shown as a histogram.

Combined

This kind of tracking curve combines both the cumulative and incremental types on one curve. For those who are used to seeing such curves, this format can be quite useful. However, most other people find it somewhat confusing.

Differential

This tracks the deviation of the dependent variable from a reference value (see Figure 7.8B). The deviation is usually considered to be acceptable if it is within predetermined limits. Trends which indicate that the final value will be outside the limits are a signal that a problem is developing.

Setting Up Tracking Curves

Step 1: Select the Dependent Variables to be Controlled

Each company, and often each project, has certain project-control variables which are particularly important. Typical variables are shown in "The Dependent Variable," earlier in this chapter.

Step 2: Select the Dependent Variables

Typical variables are shown in "The Independent Variable," earlier in this chapter.

Step 3: Select the Format to Be Used

See "What is a Tracking Curve?" earlier in this chapter.

Step 4: Draw the Reference Curve

It is evident that the successful use of a tracking curve depends on the reference curve's being well thought-out and realistic. If it is not, the variations between planned and actual values will be useless as a

tool for forecasting or control. Many users of tracking curves draw
the reference curve by using judgements or standard curve shapes
(such as the "S"). For small projects, this is unsatisfactory.

The project model (or budget plan) described in "The Budget Plan
Provides the Basis for Control," provides the basis of the reference
curve. Because it defines the work to be done, the resources required, and the direct man-hours to be expended in each segment of
time, it can be used to derive the reference curves for direct man-
hours vs. time, physical progress vs. time, total man-hours vs. time,
expenditures vs. time, etc. The progress vs. time curve can then be
used to plot other variables against progress.

In deriving the reference curve of progress vs. time from the project model, we can assume that the productivity is constant with time,
or that it varies in a predetermined way. Many companies involved in
large projects use a productivity trend curve that assumes lower-than-
average productivity during the acceleration and deceleration stages,
due to "learning curve" and other effects. However, since small projects, in general, do not have a significant acceleration or deceleration
stage, the planned productivity can generally be assumed to be constant. Physical progress will then be achieved over time at the same
rate at which the direct man-hours are expended.

As described in "Fundamentals of CPM," in Chapter 3, noncritical
activities in the project can be scheduled as if they will start on their
early-start date, their late-start date, or at some time in between.
Whatever assumption is made will, of course, affect the planned progress curve: if all activities start on their early-start date, progress
will be made more quickly than if they start later. Interestingly, although it is rare for all activities on a project to start on the early-
start date, many projects are scheduled as if they will. As a result,
progress then seems to lag behind schedule. It is very useful, therefore, to construct progress curves in which the progress for both
early-start and late-start assumptions are shown. The actual progress
will, hopefully, fall between these two extremes. An example of such
a curve is provided in Figure 7.7 which shows both early-start and
late-start schedule.

Sample Tracking Curves

Figure 7.8 shows some sample tracking curves of the cumulative,
incremented, and differential type.

Using the Tracking Curve for Forecasting and Control

Identifying Variances in Work to Date

In Figure 7.2, we see that the key to effective control lies, to a great
extent, in the process of data analysis. Given that we have captured

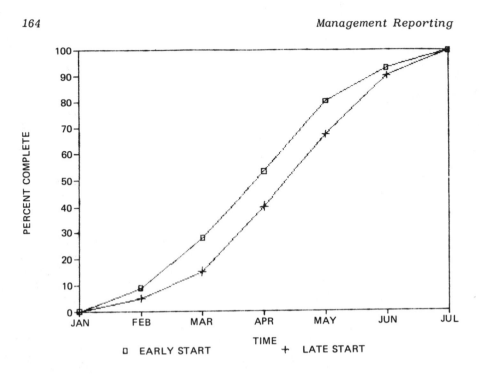

FIGURE 7.7 Cumulative progress curve showing start-date variations.

data which represents which is actually happening on the project, we must now use that data to figure out:

What has happened up to now
What is going to happen if things continue as they are now
What problems might be causing things to go wrong
What corrective action might be taken to avoid or correct those
    problems

This information is then presented in management reports, with the intention that management (that is, whoever has the appropriate authority) take the necessary action.

To answer these questions using tracking curves, it is often helpful to look at two or more tracking curves together. For example:

If:
    Progress: is below planned progress to date
    Man-hours: are below planned manhours to date

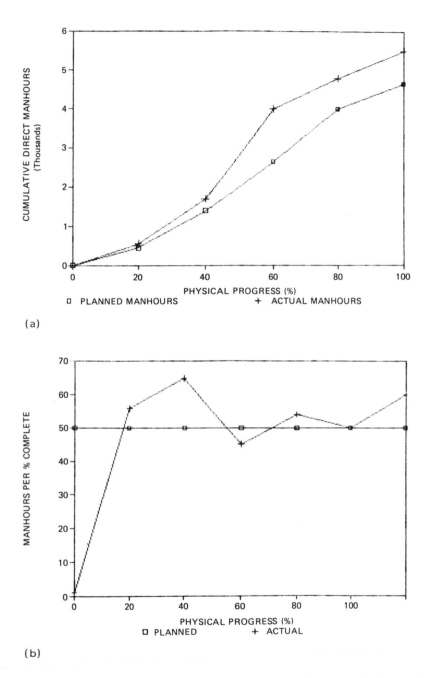

(a)

(b)

FIGURE 7.8 (a) Cumulative man-hours as a function of physical prog-ress. (b) Direct man-hours spent per % of physical progress. (c) Direct man-hours vs. planned man-hours, as a % of physical progress. (d) Planned vs. actual productivity, as a % of physical activity.

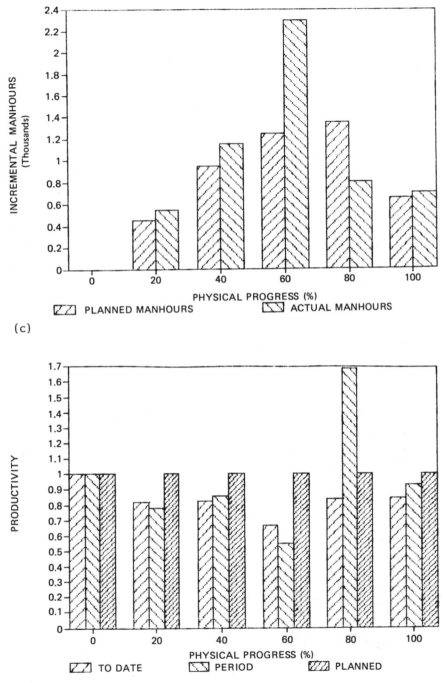

(c)

(d)

```
                              STATUS
                              ======
CONTROL PARAMETER I  AHEAD  ION TARGETI BEHIND   I
------------------------------------------------------
PHYSICAL PROGRESS I          I          I          I
                  I          I          I          I
-----------------I----------I----------I----------I
DIRECT MANHOURS   I          I          I          I
SPENT             I          I          I          I
-----------------I----------I----------I----------I
SCOPE OF WORK     I          I          I          I
                  I          I          I          I
-----------------I----------I----------I----------I
PRODUCTIVITY      I          I          I          I
                  I          I          I          I
-----------------I----------I----------I----------I
AVAILABILITY OF   I          I          I          I
RESOURCES         I          I          I          I
         MATERIAL I          I          I          I
---------I----------I----------I----------I
         LABOR    I          I          I          I
---------I----------I----------I----------I
         EQUIPMENTI          I          I          I
---------I----------I----------I----------I
         DRAWINGS I          I          I          I
---------I----------I----------I----------I
         INDIRECTSI          I          I          I
                  I          I          I          I
-----------------I----------I----------I----------I
MILESTONES        I          I          I          I
ACHIEVED TO DATE  I          I          I          I
-----------------I----------I----------I----------I
EFFECT OF EXTERNALI          I          I          I
FACTORS           I          I          I          I
-----------------I----------I----------I----------I
```

FIGURE 7.9 Project-control checklist.

We might conclude that inadequate manpower resulted in lack of
   progress
If:
   Progress: is below planned progress to date
   Man-hours: are per plan
We might conclude that productivity is below expectations

In fact, there are many combinations of project-control variations, and
each one can tell a very different story. In analyzing project data,
the project engineer often has to be a detective, using the data as
"clues" which, when taken together, form a picture of what is really
going on.

Checklists are a good way to find our way through the maze of project data. We can begin by using a checklist such as that shown on
Figure 7.9. The checklist shows that we should generally look at the
project-control parameters, in the sequence shown, ask whether each
control parameter is the same as, greater than, or less than planned.
The parameters are as follows:

Progress (i.e., physical % complete): Is progress ahead of schedule,
   on schedule, or behind schedule?
Man-hours (i.e., man-hours spent to date to achieve the given
   progress): Are the man-hours spent above, equal to, or below
   the planned man-hours?
Scope of work (i.e., quantities done to date and left to do): Is the
   quantity of work done to date greater-than, equal to, or less
   than the planned scope of work? Is the total quantity of work in
   the project greater than, equal to, or less than that which was
   planned?
Productivity (i.e., ratio of actual to planned man-hours): Is productivity greater than, equal to, or less than planned?
Resource availability (e.g., materials, labor, equipment): Has labor
   been supplied to the job in the quantities and disciplines as
   planned? Has material been provided on schedule? Have approved
   construction drawings been provided on schedule? Has the necessary construction equipment been provided as planned? Have
   overhead staff and facilities been provided as planned?
Schedule milestones: Have major schedule milestones to date been
   met?
External Factors (e.g., weather, strikes): Have external factors
   had any influence on the project to date? Are external factors
   expected to influence the work left to do?

To determine the current status of our project, then, we need to look
at the answer to each question individually, and also all the questions
together. The tracking curves should be designed to help a project
engineer complete that process quickly, by helping to identify patterns
of information.

Forecasting Future Performance

Cost and schedule forecasting is one area of project control on which
the philosophy and judgement of the individuals and organizations involved can have a major impact. Some companies and individuals prefer to forecast that the final cost and schedule will be identical to the
budget plan, in the hope that any delays or overruns which are currently foreseen will be offset by future lucky breaks. This tendency
to cover-up potential overruns or delays has the effect of reducing

project-control effectiveness since problems requiring corrective action are not highlighted.

There is also a natural tendency to avoid criticism and punishment. The dilemma of the project engineer is often that if he forecasts a delay or overrun, there is sure to be trouble right away, but if he reports that the project is on schedule and budget, he might (or might not) get into trouble later if and when the overrun or delay actually occurs.

If a cost or schedule forecast is to be effective as a project control device, it should be defined as follows:

"This cost (or schedule) forecast represents the final actual cost (or schedule) that will be experienced *if no action is taken to correct the problems or adverse trends identified to date.*"

In other words, the forecast is not what we think will actually happen, it is what we think will actually happen if no changes are made. Therefore, a tracking curve can be used to extrapolate data trends to forecast what the curve of actual data will look like. The extrapolation is usually based on:

1. The variations defined by the plot of actual data to date (e.g., progress per month is 75% of the planned progress rate)
2. The shape of the reference curve (e.g., an S curve)
3. The slope of the trend curve through the actual data (e.g., a "flatter" slope)

Tracking curves are a useful way to assure that the possibility of delays and/or cost overruns cannot be easily hidden. The data points reflecting actual performance will generally follow one of several patterns, as described in the following:

No significant deviation from the plan

In this case, we have good grounds for forecasting that the dependent variable will follow the reference curve and that the project will finish on time and within the budget.

Clear adverse trend

When actual data points show a clear adverse trend, such as progress being consistently below expectations, it is a clear signal that a delay and/or overrun can be expected if nothing is done. We can expect that the actual values of the dependent variable will follow the shape of the reference curve, but that the slope will reflect the trend established to date. So, for example, if we are a week behind schedule because we are making less progress than planned each week, the cumulative effect is that we will be much more than a week late at the end of the job.

Clear favorable trend

If the actual data indicates a clear positive trend, it may indicate that a cost underrun or schedule improvement is possible. However, such favorable results should be reported only at such time as the reason for the favorable results to date can be identified, and it an be expected that the favorable trend will continue.

Scattered data with no clear trend

Tracking-curve data often shows a degree of scatter which is clearly beyond the expected variation. Not only does such data make it impossible to predict trends and make forecasts, but it indicates a lack of credibility in the project-control methods and data. If such data is experienced, it almost certainly indicates that something is wrong with the data collection and analysis method. One frequent problem area is in the definition of terms. For example, it may be that indirect man-hours are being reported as direct, or the progress-measurement system may contain some anomalies. In any case, the problem should be identified and rectified.

It should be noted that some project-control variables, such as productivity, may well experience a scatter pattern, and not exhibit a clear trend.

THE HEIRARCHY OF REPORTING

For management attention to be focused effectively on those areas requiring action, it is essential that reports be concise and to the point. As shown in Figure 2.5, the level of detail required by management is a function of the level of the user in the organization: the higher up in the organization the user is, the less detail is important.

It is possible, therefore, to visualize a family of reports, all generated from the same project model, in which the level of detail is matched to the level in the organization of the person receiving it. Figures 7.10 through 7.15 show examples of this heirarchy of reporting for progress and cost control. The first report (Figure 7.10) shows only summary information. It is intended for identification of those projects which are experiencing or might experience delays and/or overruns. This report might be of interest to the engineering manager who is responsible for the department that handles all the small projects. The second-level schedule report (Figure 7.11) provides additional planning information such as the start dates and man-hours of each project. It is still, however, a summary report on which a number of projects can be displayed. This level of detail might be appropriate for a supervisor of several project engineers, each of whom handles several projects.

PROJECT NO. NAME	BUDGET VALUE $	CURRENT FORECAST $	OVER RUN UNDER RUN $	%	PLANNED COMPLETION M/D/Y	FORECAST COMPLETION M/D/Y	AHEAD BEHIND DAYS
123 PUMP SERVICING	$3,600.00	$3,500.00	($100.00)	-3%	MAY 1,85	MAY 5,85	4
132 HT. EXCH BUNDLES	$5,400.00	$5,600.00	$200.00	4%	MAY 30,85	MAY 30,85	0
143 ROAD REPAIR	$12,400.00	$12,400.00	$0.00	0%	JUNE 15,85	JUN 15,85	0
146 REROUTE PIPING	$8,700.00	$9,500.00	$800.00	9%	JUNE 25,85	JUN 30,85	5
153 NEW SWITCHGEAR	$26,900.00	$24,500.00	($2,400.00)	-9%	AUG 2, 85	AUG 15,85	13
177 LAB BUILDING ROOF	$3,500.00	$3,500.00	$0.00	0%	JUNE 16,85	JUN 20,85	4
187 REPLACE VALVES	$18,700.00	$20,100.00	$1,400.00	7%	JULY 28,85	JULY 20,85	-8
TOTAL	$79,200.00	$79,100.00	($100.00)	0%			

FIGURE 7.10 Small-project summary report.

PROJECT NO. NAME	COMPLETION DATE			START DATE			TOTAL MANHOURS			PROGRESS TO DATE		
	PLAN	FORECAST	VAR. DAYS	PLAN	ACTUAL	VAR. DAYS	PLAN	F-CAST	VAR	PLAN %	ACT. %	VAR. %
123 PUMP SERVICING	MAY 1	MAY 5	4	APR 1	APR 3	2	120	150	30	90	86	-4
132 HT. EXCH BUNDLES	MAY 3	MAY 30	0	MAR 1	MAR 15	0	180	180	0	30	30	0
143 ROAD REPAIR	JUN 1	JUN 15	0	FEB 2	MAR 1	2	400	420	20	55	55	0
146 REROUTE PIPING	JUN 2	JUN 30	5	MAR 2	MAR 28	3	350	380	30	38	33	-5
153 NEW SWITCHGEAR	AUG 2	AUG 15	13	APR 1	APR 15	0	650	650	0	15	7	-8
177 LAB ROOF	JUN 1	JUN 20	4	APR 1	APR 1	0	150	160	10	42	39	-3
187 REPLACE VALVES	JULY 1	JULY 20,	-8	MAY 1	MAY 4,8	3	350	320	-30	5	9	4

FIGURE 7.11 Small-project schedule-summary report.

DATE: MAY 4, 1985

PROJECT NO. NAME	BUDGET VALUE $	APPROVED CHANGES $	CURRENT ESTIMATE $	CURRENT FORECAST $	OVER RUN UNDER RUN $	%	VALUE OF WORK DONE PLANNED $	ACTUAL $	VARIANCE %
123 PUMP SERVICING	$3,800.00	$150.00	$3,750.00	$3,500.00	($250.00)	-7%	3240	3010	-230
132 HT. EXCH BUNDLES	$5,400.00	$280.00	$5,680.00	$5,600.00	($80.00)	-1%	1620	1680	60
143 ROAD REPAIR	$12,400.00	$100.00	$12,500.00	$12,400.00	($100.00)	-1%	6820	6820	0
146 REROUTE PIPING	$8,700.00	$680.00	$9,380.00	$9,500.00	$120.00	1%	3306	3135	-171
153 NEW SWITCHGEAR	$26,900.00	($950.00)	$25,950.00	$24,500.00	($1,450.00)	-5%	4035	1715	-2320
177 LAB BUILDING ROOF	$3,500.00	$0.00	$3,500.00	$3,500.00	$0.00	0%	1,470	1365	-105
187 REPLACE VALVES	$18,700.00	$1,200.00	$19,900.00	$20,100.00	$200.00	1%	935	1809	874
TOTAL	$79,200.00	$1,460.00	$80,660.00	$79,100.00	($1,560.00)	-2%			

FIGURE 7.12 Small-project cost-summary report.

WORKWEEK 7DAYS   9 HOURS PER DAY

DATE: APR 15,85

ACTIV. NO.	ACTIVITY NAME	START DATE PLAN	ACTUAL	DURATION PLAN	F.CAST/ACTUAL	FINISH DATE PLAN	F.CAST	MANHOURS PLAN	F.CAST TOTAL	PROGRESS TO DATE % PLAN	% ACTUAL	VARIANCES MHRS.	PROG. %	START DAYS	DUR. DAYS	FIN. DAYS
1208	EXCAVATE FOUNDATIONS	APR 5	APR 10	3	3	APR 8	APR 13	288	300	100	100	4	0	5	0	5
2208	POUR FOUNDATIONS	APR 8	APR 13	10	9	APR 18	APR 22	480	460	90	15	-4	-75	5	-1	4
1211	SET HEAT EXCHANGERS	APR 18	APR 22	2	3	APR 20	APR 25	96	130	0	0	35	0	4	1	5
1213	FABRICATE PIPING	APR 4	APR 4	14	15	APR 18	APR 19	680	800	40	65	18	25	0	1	1
2213	ERECT PIPING	APR 18	APR 19	6	5	APR 24	APR 24	864	750	0	0	-13	0	1	-1	0
3213	FLUSH & TEST PIPING	APR 24	APR 24	1	1	APR 25	APR 25	48	50	0	0	4	0	0	0	0
1218	FABRICATE PIPERACK	APR 3	APR 9	5	8	APR 8	APR 17	300	420	100	80	40	-20	6	3	9
2218	ERECT PIPERACK	APR 8	APR 17	2	3	APR 10	APR 20	120	150	100	0	25	-100	9	1	10
1222	INSTALL INSTRUMENTS	APR 12	APR 12	3	2	APR 15	APR 14	180	130	100	100	-28	0	0	-1	-1
2222	TEST & COMMISSION	APR 25	APR 25	2	3	APR 27	APR 28	144	190	0	0	32	0	0	1	1
1225	STARTUP	APR 27	APR 28	1	1	APR 28	APR 29	144	150	0	0	4	0	1	0	1
	TOTAL			49	53			3344	3530							4

FIGURE 7.13 Small-project detailed schedule report.

COST CODE	DESCRIPTION	BUDGET MAT'L	BUDGET LABOR	BUDGET TOTAL	FORECAST MAT'L	FORECAST LABOR	FORECAST TOTAL	VAR MAT'L $	VAR MAT'L %	VAR LABOR $	VAR LABOR %	VAR TOTAL $	VAR TOTAL %
	COST SUMMARY												
208	FOUNDATIONS	8500	15400	23900	8800	16500	25300	300	3.5	1100	7.1	1400	5.9
211	HEAT EXCHANGERS	85000	12500	97500	82600	12700	95300	-2400	-2.8	200	1.6	-2200	-2.3
213	PIPING	78100	67000	145100	81500	68500	150000	3400	4.4	1500	2.2	4900	3.4
218	STRUCT. STEEL	45600	13400	59000	48700	15600	64300	3100	6.8	2200	16.4	5300	9.0
222	INSTRUMENTATION	23500	3400	26900	25000	3400	28400	1500	6.4	0	0.0	1500	5.6
	TOTAL	240700	111700	352400	246600	116700	363300	5900	2.5	5000	4.5	10900	3.1
	COST DETAILS												
208	FOUNDATIONS												
	F101	3400	6160	9560	3520	6600	10120	120	3.5	440	7.1	560	5.9
	F102	2550	4620	7170	2640	4950	7590	90	3.5	330	7.1	420	5.9
	F013	2550	4620	7170	2640	4950	7590	90	3.5	330	7.1	420	5.9
211	HEAT EXCHANGERS												
	E101	42500	6250	48750	41300	6000	47300	-1200	-2.8	-250	-4.0	-1450	-3.0
	E102	42500	6250	48750	41300	6700	48000	-1200	-2.8	450	7.2	-750	-1.5
213	PIPING												
	PIPE	35800	35000	70800	37300	35500	72800	1500	4.2	500	1.4	2000	2.8
	VALVES	34500	13000	47500	35500	16000	51500	1000	2.9	3000	23.1	4000	8.4
	FITTINGS	7800	19000	26800	8700	17000	25700	900	11.5	-2000	-10.5	-1100	-4.1
218	STRUCT. STEEL												
	SUPPORT BEAMS	20000	8400	28400	22500	9000	31500	2500	12.5	600	7.1	3100	10.9
	PLATES & SHAPES	0	0		0	0		0		0		0	
	LIGHT STEEL	25600	5000	30600	26200	6600	32800	600	2.3	1600	32.0	2200	7.2
222	INSTRUMENTATION												
	CONTROL VALVES	12300	600	12900	16500	700	17200	4200	34.1	100	16.7	4300	33.3
	GAGES & SENSORS	4700	1100	5800	2400	1100	3500	-2300	-48.9	0	0.0	-2300	-39.7
	WIRING	3200	1100	4300	3000	1100	4100	-200	-6.3	0	0.0	-200	-4.7
	DISPLAYS	3300	600	3900	3100	500	3600	-200	-6.1	-100	-16.7	-300	-7.7
	TOTAL	240700	111700	352400	246600	116700	363300	5900	2.5	5000	4.5	10900	3.1

FIGURE 7.14 Small project-detailed cost report.

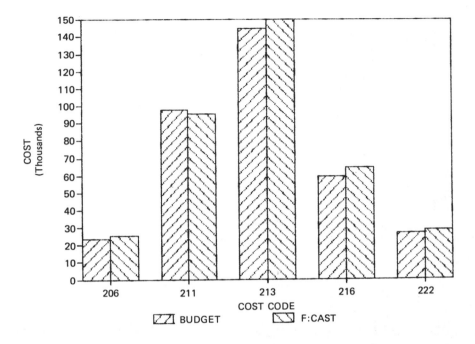

FIGURE 7.15 Budget vs. Forecast costs, by cost code.

Similarly, the second-level cost report (Figure 7.12) would be used by
the same supervisor: it also provides summary information on a number
of projects. Note that the "various" columns permit management by exception. By scanning those columns, the reader's attention is
immediately focussed on the problem areas.

The third level of report is for the project engineer. The purpose
of these reports is to pinpoint those specific areas of a given project
which are causing the overrun or delay. Figure 7.13 provides a sample
schedule analysis report. This report would, in general, be prepared
for one project at a time. It shows the kind of detailed information that
enables the project engineer to identify the specific activities which
have problems and, by deduction, the nature of the problem. Similarly,
Figure 7.14 provides the same level of information regarding costs, and
this information is summarized graphically in Figure 7.15.

The project model is most helpful in preparing these reports. If we
think of the project model as a heirarchy of information, the lowest
level of the heirarchy shows the maximum amount of detail. That level
is the level at which the work is actually done. For example, if company workforces are performing the construction work, the management level at the "workface" is the foreman. Below him are the people

whose man-hours are considered direct and whose efforts result in measurable physical progress. To the foreman, the project consists of those activities for which he is responsible. The resources are the people who work for him, the materials they install, the engineering drawings he needs to do the work, and the construction equipment he uses. His work is planned and recorded using timesheets, task-sheets, job-cards, work-orders, or some other detailed description of specific work scope. Progress is measured at this level of detail, as are manhours. This level of detail is the level of detail of the project model.

At the next level in the small-project management heirarchy, we might find the project engineer (or the field superintendent, construction manager, site engineer, etc.) Regardless of his title, the person at this level has a perspective which encompasses the entire project's scope and schedule. The information required to manage at this level can be derived easily by summarizing information from the detailed level.

The next level of detail for the small project is generally one in which several projects are considered at once. These reports can also be prepared by aggregation from the project models. For those using the higher-level reports with summary-type information, the additional detail available at the lower level reports is used for problem identification, analysis, and resolution. It can be thought of as an "audit trail": there when needed but not evident when not needed.

The essential points about the heirarchy of reporting are these:

1. The perspective which is essential to effective project control can only be attained when excessive detail is eliminated from management reports.
2. The level of detail should match the level of the person using the data.
3. Reporting should facilitate management by exception.
4. Detailed information should be available and used for problem identification and resolution.

USING REPORTS TO MANAGE THE WORK OF OTHERS

The Problem of Objective Reporting

The classic problem of project control stems from the following situation:

The project engineer (or manager) is not in a good position to collect and analyze detailed information, so therefore he must rely on the contractor to do it.

The contractor (or subcontractor) cannot be blamed for having a
natural aversion to reporting "bad news," particularly if it reflects
poorly on his own performance.

To put it another, more general, way, progress reports are intended
to report on the performance of the people doing the work to those who
pay for it. However, the only people who are in a position to collect
data and prepare the progress reports are those whose performance is
being measured and reported. It is only natural that, under those
circumstances, there is apt to be a certain lack of objectivity in reporting, particularly when those preparing the report are certain that,
if the report contains unfavorable information, a lot of trouble will result. It is easy, under those circumstances, to convince oneself that
there is no need to report adverse trends because something good may
happen soon which will offset those cost overruns and schedule delays.
Yet, there can be no question that timely, objective reporting is
essential to project control. How then can we assure that the natural
bias against reporting accurately is overcome? The solution is to design
and implement a project-control system which leaves no room for
"cheating." This requires that the basic project data, such as man-
hours and physical progress, be captured, analyzed, and reported in
a rigorous way. Computerized systems, of course, are an aid in this
but are not essential. Usually, one finds that the data required for
payment of an invoice, such as man-hours or progress, is carefully
monitored and reported and therefore forms a good basis for control.

Using the Heirarchy of Networks

If we can break a project down into its component packages of work,
and then manage and control those packages, we generally find that
the smaller a work package is, the easier it is to control. Therefore,
the use of "hammocks" and "subnetworks," as described in "Definition
of Planning Terminology," in Chapter 3, (see Figures 3.3 and 3.4),
enables the project engineer to measure progress and man-hours at a
lower level of detail than he would normally work in himself. For
example, progress measured at the subcontract level can be aggregated
to reflect progress in the major subprojects which can, in turn, be
aggregated to reflect progress in the entire project.
Progress measurement methods are discussed in Chapter 8. On small
projects, where formal progress measurement systems may not be available, judgement is often the only way to measure progress. This method, discussed more fully in Chapter 8, requires that the work packages
on which progress is to be measured be small enough that the judgements are likely to be reasonable. The larger the work package, the
more difficult it is to measure progress by judgement.

CAPTURING DATA

In the operating-plant environment in which many small projects take place, it is unlikely that the project engineer will be in a position to force substantial changes to existing practices and procedures, so that he can have the project control data he needs. It is important, then, to carefully identify that data which is available from existing procedures. Project data can then be selectively captured by tapping into the flow of data which already exists.

Before project-control methods, systems, and procedures are designed, a careful review should be made of all existing practices and procedures. The specific documents which provide project data such as man-hours, costs, resource status, etc., should be identified. The data on each document which is required by the project engineer should then be identified. This data should be kept to a minimum to allow it to be captured quickly. The means by which the project engineer will store the data and use it to update the project model is then defined: it might be by manual or computerized input, or even perhaps by a transfer of computer files. Finally, the data needed for project control which is not available from existing sources is identified, and procedures established for capturing it.

CHAPTER SUMMARY

In this chapter we discussed the importance of management reporting to the project-control function. We described the use of tracking curves as an appropriate forecasting and control technique for small projects, the use of the project model as a basis for control, and heirarchy of networks that can be built around the project model to provide timely information at each level in the organization.

8

Measurement of Physical Progress

THE IMPORTANCE OF PROGRESS MEASUREMENT

Project control depends, to a very great extent, on the effectiveness of the methods used to measure and report physical progress. Even on small projects, a progress-measurement method is essential as it defines "where we are" such that we can compare that with "where we planned to be," and thereby identify areas needing attention.

Small projects, because of their size and other aspects of the small-project environment, often experience ineffective project control because of the lack of progress measurement. Proper progress measurement is often thought to be simply too much time and effort for the small project, but this need not be the case. A progress-measurement method can be defined once, and then implemented on all the small projects. In this chapter we explore the basic principles of progress measurement as they apply to projects of all sizes, and then discuss various shortcut techniques that can be applied.

PRINCIPLES OF PROGRESS MEASUREMENT

Elements of an Effective Progress-Measurement Method

An effective progress-measurement method should do or be the following:

Provide a measure of the physical quantities of work done
Provide a measure of the current total scope of work of the project

Express the work done as a percentage of the total current scope of work

Be unbiased (i.e., it should not be significantly affected by optimism, pessimism, or bias)

Be realistic (i.e., it should reflect the many hard-to-measure items of work that, individually, may be small but which, collectively, contribute significantly to the scope of work)

Be agreed upon (by those whose performance is being measured and by those who are doing the measuring)

Be fair (to those who are doing the work and to those who are paying for it)

Be efficient (i.e., it should not require excessive time to collect, analyze, and present data)

Be well documented (to assure consistency)

Be independent of actual man-hours (i.e., it should measure actual physical work done without regard to the number of man-hours spent)

Basic Steps in Progress Measurement

The basic steps that are required to set up and implement a progress-measurement method are described below. Each step will be illustrated in "The Earned Value Method of Progress Measurement," later in this chapter.

Step 1: Divide Scope of Work into Packages for Control

Most project-management techniques are based on breaking a project down into components or packages of work, the idea being that each package of work be manageable, and appropriately sized to the level of the person or company handling it. Often a "work package" is assigned to a specific individual, organization, function, or contractor. Examples of work packages are:

An activity on a network plan
A work package in a "work breakdown structure"
A contract or subcontract
A "work order," task-sheet," or "job-card"
A variation to a contract
The work of one craft or discipline
A specified quantity of work
The work in a given geographical area
The work covered by a certain code in the code of accounts

For progress measurement, it doesn't really matter what form the work packages take, so long as *the total amount of work in all packages equals the total scope of work of the project.* Naturally, it is preferable that progress be measured against network activities so that the progress to date can be used as the basis for a schedule forecast, but this is not essential.

The scope of work may actually be divided several times until sufficient detail is reached. The work packages should also be compatible with the way man-hours are collected, to facilitate productivity measurements.

Step 2: Establish the Standard Work Unit

If progress in diverse activities is to be measured and then aggregated, it is helpful to have a unit of measure which can be applied to any activity, regardless of the type of work being done. For example, we could not add progress in the form of cubic yards of concrete poured, to progress in the form of linear feet of electrical cable installed, yet we would like to be able to calculate the net progress for both civil and electrical work. To do so, we define a standard work unit with which all progress can be expressed. The most frequently used standard work unit is the "earned man-hour." This is the basis of the "earned-value" system of progress measurement which will be described later in this chapter. The advantage of the earned man-hour approach is that earned man-hours can be easily compared with actual man-hours to give an indication of productivity.

Step 3: Define the "Yardstick" for
Measuring Progress

This step consists of defining, *in advance*, what amount of work or milestones constitutes what % complete. This can be done by judgement, experience, or by an allocation of estimated man-hours. It consists of identifying the physical results of the direct labor in the project, as well as the milestones which are required to complete the work on each item.

Step 4: Define the Method For
Aggregating Progress

This step consists of defining the means by which progress in one category will be added to progress in other categories to arrive at the overall % complete.

Step 5: Define Who Will Measure Progress, and
How Often

The individuals or functions who will actually do the progress measurement should be defined as part of the procedure. It should be clearly understood who will do the measuring, what data will be required, how often it will be required, who will collect and analyze the data, to whom it will be presented, etc.

Step 6: Agree on the Progress-Measurement Method Prior to the Start of Work

Progress measurement is, at best, a combination of numerical analysis and judgement. Because the individual or organization whose performance is being measured must, in general, provide much of the information necessary to measure his performance, it is evident that good cooperation and communication is essential if the system is to work. For this reason, it is important that the method for measuring progress be agreed upon prior to the start of work. Both those being measured and those doing the measuring must agree that the method for measuring progress is fair.

ESTABLISHING A PROGRESS-MEASUREMENT METHOD FOR SMALL PROJECTS

Although small projects, taken alone, often seem to be too small to justify a formal progress-measurement system, the fact is that, once a progress-measurement system is established, it can be used over and over again on all small projects. Just as in planning, contracting, cost estimating, and other project-management functions, the benefits of increased consistency, effectiveness, and efficiency far outweigh the effort required to develop and implement such methods.

The progress-measurement system should cover the three major functions of a project: design, procurement, and construction. (Of course, not all projects contain all three functions.) The progress-measurement system should cover all work that is considered direct, and should focus, whenever possible, on the *physical results* of the work, instead of the effort required. For example, it is often said that it is difficult to measure progress on work done by engineers, scientists, programmers, and other "white-collar" persons. In such cases, we can ask ourselves, "what physical result is produced by the efforts of this person?" Perhaps it is a drawing, or a report, computer program, design specification, bill of materials, purchase order, contract, etc. When the items which represent the physical work are identified, it is a simple matter to identify the major milestones involved, and assign %-complete values to them.

THE EARNED VALUE METHOD OF PROGRESS MEASUREMENT

The best-known method of progress measurement is the earned value method in which the work done is expressed in terms of work units earned. In most cases, the work units are standard man-hours (also referred to as base man-hours, "norm" man-hours, or estimated man-hours). As progress is made, we "earn" man-hours, and can sum these man-hours to yield a calculation of overall progress. The earned-value method also provides a good illustration of the basic approach which will be used in all progress-measurement methods. Each step of the earned value method is illustrated in the following example.

This construction project involves subcontracts for civil, architectural, mechanical, electrical, and steelwork.

Step 1: Divide the scope of work into packages for control

Let the major packages be the major subcontracts:

   a. Civil
   b. Architectural
   c. Mechanical
   d. Electrical
   e. Steelwork

Let the component packages be as follows:

   a. Civil subcontract
      1. Site clearance and grading
      2. Foundations
      3. Paving
   b. Architectural subcontract
      1. Walls
      2. Roof
      3. Glass
      4. Insulation, paint, etc.
   c. Mechanical subcontract
      1. HVAC (heating, ventilating, and air conditioning)
      2. Potable water pumps and piping
   d. Electrical subcontract
      1. Transformer
      2. Switchgear
      3. Cable and conduit
      4. Environmental control system
   e. Steel erection subcontract
      1. Heavy steel
      2. Light steel

Step 2: Define the standard work unit

The standard work unit will be standard man-hours (man-hours at the productivity assumed for the base estimate).

Step 3: Define the yardstick for measuring progress.

For example:

Component	Work package	Percent earned when complete
a. Civil subcontract		
1. Site clearance and grading		
	grub and clear	25
	cut and fill	35
	rough grade	20
	final grade	20
		100
2. Foundations		
	excavation	25
	forms	15
	reinforcing bars	20
	pour and cure	25
	strip forms and backfill	15
		100
3. Paving		
	grading	45
	base course	30
	final course	25
		100
b. Architectural subcontract		
c. Mechanical subcontract		
d. Electrical subcontract		
e. Steel erection subcontract		

The yardsticks for the other subcontracts are defined in exactly the same way. To establish the percent earned at the completion of each work package, we could have used judgement or we could have estimated the relative man-hours to be spent on each work package and set the percent earned equal to the percent of man-hours.

Step 4:   Define the method for aggregating progress

Aggregating process at the subcontract level

Subcontract	Component	Percent complete	Standard man-hours	Earned man-hours
a. Civil				
	1. site clearance and grading	23	1500	345
	2. foundations	11	2450	270
	3. paving	5	900	45
Total man-hours, this subcontract:			4850	660

Net % complete = earned man-hours/standard man-hours
                       = 660/4850 = 13.6%.

Progress on each subcontract is calculated in the same way.

Aggregating progress at the project level

Subcontract	Total standard manhours	Percent of total MH	Subcontract complete (%)	Contribution to project progress (%)
a. Civil	4850	40.8	13.6	5.6
b. Architectural	2650	22.3	8.2	1.8
c. Mechanical	1500	12.6	5.0	0.6
d. Electrical	1650	13.8	7.0	1.0
e. Steel erection	1250	10.5	12.5	1.3
	11900	100		10.3

The construction work is 10.3% complete overall.

Step 5:   Define who will measure progress and how often

Progress will be measured weekly by each subcontractor using yard-sticks agreed upon with the prime contractor and the client. Progress measurement will be reviewed by the prime contractor and the client.

Step 6:   Agree on the progress-measurement method prior to the start of work

The procedures for measuring progress and the responsibilities of each partly are defined in an attachment to the basic contracts. These procedures should be thoroughly discussed and agreed upon prior to the start of work.

MEASUREMENT OF PROGRESS IN
ENGINEERING WORK

A common misconception with regard to white-collar work activities, such as design engineering, drafting, programming, research, etc., is that its creative nature precludes effective progress measurement and control. This argument is, naturally, often put forward by those whose progress is to be measured. In fact, progress measurement on such work activities can be conducted in the same way that it is on construction.

In general, work which is worth measuring will result in something physical. For example:

> *Engineering work* results in equipment specifications, bills of material, design reports, process flow diagrams, etc.
> *Design and drafting work* results in drawings (layouts, general arrangements, piping and instrument drawings, electrical one-line drawings, etc.)
> *Inspection and testing work* results in a "certificate of fitness" or equivalent document

If we follow the sequence of activities described in "Establishing a Progress-Measurement Method for Small Projects" (earlier in this chapter), we might find that a progress-measurement system for engineering would look something like the following.

Step 1: Divide the scope of work into packages for control

Let the major packages be the engineering departments:

a. Civil and structural
b. Mechanical
c. Process
d. Electrical

Let the component packages be as follows:

a. Civil and structural department
  1. general arrangements, plot plans, layout, and excavation drawings
  2. heavy structural steel arrangement drawings
  3. excavation drawings
  4. underground piping and electrical drawings
  5. foundation design analyses
  6. structural design analyses
  7. civil and structural detail drawings

    b. Mechanical department
       1. HVAC equipment design analysis and specification
       2. HVAC layout
       3. pressure-vessel design analysis and specification
       4. rotating-equipment design analysis and specification
       5. fired-heaters design analysis and specification
       6. heat-exchanger design analysis and specification
    c. Process department
       1. Process analysis
       2. heat and material balance calculations
       3. process-flow diagrams
       4. piping and instrumentation diagrams
    d. Electrical department
       1. load analysis
       2. equipment sizing and specification
       3. one-line drawings
       4. electrical detail drawings

Step 2: Define the standard work unit

The standard work unit will be standard man-hours, i.e., man-hours at the productivity assumed for the base estimate.

Step 3: Define the yardstick for measuring progress

As in the case of construction work, detailed man-hour estimates for each work package can be used as guidelines for assigning progress. Judgement, based on milestones and completion of key activities, can also be used. For example:

Preparation of design drawings

Work package	Percent earned when complete
Preliminary sketches	10
Initial layout and discipline review	35
Incorporate vendor drawings	45
Interdiscipline review and revisions	50
Design complete approved by client	75
Changes and revisions	95
Approval for construction	100

Another example involves design analysis:

Work package	Percent earned when complete
Kickoff meetings and job scoping	5
Initial hand calculations	15
Preliminary computer analysis	35
Review of preliminary design	40
Secondary computer analysis	60
Interim review and revisions	75
Final analysis	85
Preparation of specification	95
Vendor followup	100

Step 4A: Aggregating progress at department level

Discipline	Activity	Percent complete	Standard man-hours	Earned man-hours
a. Civil/structural				
	layouts	65	360	234
	foundation drawings	75	200	150
	structural drawings	40	120	48
	underground drawings	45	140	63
	foundation design	65	220	143
	steel design	75	185	139
	detail drawings	35	110	39
Total man-hours, this department			1335	816

net % complete = earned man-hours/standard man-hours
= 816/1335 = 61%.

Progress in each department is calculated in exactly the same way.

Step 4B: Aggregating progress for all engineering work

Department	Total standard man-hours	Percent of total MH	Department % complete	Percent contribution to project progress
Civil/structural	1335	38	61	23
Mechanical	750	22	42	9

Department	Total standard man-hours	Percent of total MH	Department % complete	Percent contribution to project progress
Process	550	16	50	8
Electrical	850	24	32	8
	3485	100		48

The engineering work is 48% complete overall.

Step 5: Define who will measure progress and how often.

Progress will be measured weekly by each department head and will be reviewed with the project engineer as well as the engineering-division manager.

MEASUREMENT OF PROGRESS IN PROCUREMENT

Like engineering, procurement is a "home-office" activity whose results do not seem to lend themselves well to progress measurement. However, procurement activities are often on the critical path, as the timely delivery of materials and equipment are critical to the schedule on a small project. Therefore, the ability to manage progress in the procurement function can be of vital importance to the small (or large) project.

Like engineering, procurement work results in the production of identifiable work packages and the completion of identifiable tasks. For example:

Preparation, approval, and award of purchase orders
Traffic (i.e., shipping arrangements)
Expediting
Inspection
Liason with engineering, planning, warehousing, construction, and
other groups

Each of these activities can be broken down into discrete tasks, and %-complete values assigned just as in the above examples for construction and engineering work. For example:

Major work package is the award of purchase order for major equipment
The component packages for this work are as follows (Step 1):

1. Define general requirements (e.g., a 400 hp centrifugal compressor)

2. Screen potential bidders to identify those willing and able to bid, and establish bidders list
3. Prepare and release invitation to bid documents
4. Receive, open, and evaluate bids and prepare bid tabulation
5. Obtain authorization for purchase order
6. Conclude final negotiations and award purchase order
7. Administer revisions and payments
8. Close out purchase order

A progress measurement yardstick might then look as follows (Step 3):

Measuring progress for each purchase order

Work package	Percent earned when complete
1. Define general requirements	5
2. Establish bid list	15
3. Release invitation to bid	25
4. Bid tabulation	45
5. Authorization obtained	55
6. Award purchase order	70
7. Administer	90
8. Close out	100

Aggregating progress for all purchase orders (Step 4A):

Purchase order	Standard man-hours	Percent complete	Earned man-hours
A101	120	50%	60
A102	150	60%	90
C402	130	40%	52
D503	140	45%	63
total manhours	540		265

net % complete = earned man-hours/standard man-hours
                = 265/540 = 49%.

Aggregating progress for the procurement function (Step 4B):

Work package	Total standard man-hours	Percent of total man-hours	Package % complete	Percent contribution to project progress
Purchasing	540	45	49	22
Traffic	150	12	15	2
Expediting	160	13	17	2
Inspection	160	13	11	1
Liaison	200	17	30	5
	1210	100%		32%

The procurement function is 32% complete overall.

MEASUREMENT OF PROGRESS IN PROJECT MANAGEMENT

Probably the most difficult aspect of project work to measure physical progress in, is project management. This may not be a problem on most small projects in operating plants in which project management is considered an overhead and is not charged to the project. However, on small projects which must "stand alone," the project management cost is apt to be significant enough to justify the effort to control it.

Our approach to defining a progress-measurement system for project management is the same as for all other types of work: to break the work down into measurable tasks or products. For project management, we can define the physical results of project-management efforts as including:

Progress reports
Cost estimates
Network plans
Schedules
Contracting plans
Purchasing plans
Quality control program

We can also define key milestones for each project-management function, which can also be linked to progress. We would then assign a value for percent complete to each milestone. Progress can be assumed to be linear between the milestones. For example:

1. For the overall project-management function:
   Authorization to proceed
   Award of design-engineering contract
   Completion of conceptual design
   Placement of major purchase orders (e.g., for long-lead
   equipment)
   Award of construction contract
   Completion of engineering
   Completion of construction
2. For the planning and cost-engineering function:
   Completion and approval of the budget plan
   Preparation of each cost and schedule forecast
3. For the contract and procurement function:
   Award of major contracts
   Placement of major purchase orders
   Delivery of major equipment
   Completion of major contracts
4. For the quality-control function:
   Approval of Quality Assurance/Quality Control (QA/QC) plan
   Inspection and approval of major fabrication items
   Inspection and approval of major construction items
   System commissioning and start up

This milestone technique is discussed further in "Shortcut Techniques for Measuring Progress," later in this chapter.

PROVISION FOR CHANGES

A classic problem with progress-measurement systems is the problem of coping with changes. In most cases, if we measure progress against the original scope of work, we are apt to encounter periods in which progress is offset by increases in the scope of work. Larger projects have been known to experience negative progress in months in which a large number of changes have been approved. It has not been easy for those project engineers to explain to management how they were able to spend thousands of man-hours during the month and have −5% progress to show for it!

In the discussion of contingency presented in Chapter 10, we find that contingency is provided for in an estimate and schedule for those variations which are likely to occur but which cannot be specifically identified at the time the estimate is prepared. It is therefore un-realistic to state that we are, for example, 50% complete with a project if that 50% refers to the original scope of work which is likely to be less than the actual scope. If is far more realistic to include the amount of work covered by contingency in the progress calculation. If it turns out that this contingency amount is not used entirely, we will simply enjoy a faster-than-anticipated rate of progress in the final stages of the project (during which progress is often slow anyway). Progress should, therefore, be measured against the expected total scope of work. This can be done by adding a man-hour contingency to the standard man-hours, and tracking actual (and earned) man-hours for design changes, field changes, and other categories of work covered by contingency. Alternatively, actual progress and man-hours can be measured against the current approved scope of work, that is, the original scope plus approved changes.

SHORTCUT TECHNIQUES FOR MEASURING PROGRESS

Methods Based on Judgement

On many small projects—such as projects with short durations, or perhaps projects which are executed simultaneously with many others— a shortcut approach to progress measurement is sufficient. If used correctly and consistently, there is nothing wrong with shortcut techniques for progress measurement. Certainly, an approximate progress measurement is far better than no progress measurement at all.

In general, an approximate method for progress measurement must rely on judgement. The use of judgement, a time-honored method, has fallen into some disrepute as the marked tendency of such judgements to be optimistic has become more apparent. Judgements on progress often suffer from the fact that those whose performance is being measured can hardly be expected to be objective in making that measurement, and those who dispute the measurement have nothing more than their own opinion on which to base their arguments. However, progress measurements based on judgement can be a valid approximation if the basis for making that judgement is clearly defined and consistently applied. Once a method and guidelines are established for small projects they can, of course, be used on all projects and often end up saving time that would otherwise be spent in progress-related disputes. Some methods for using judgement in approximating progress are described below.

Using the Network Activities for Progress Measurement

If a project is represented by a network, there will be some activities that are complete, some that are in progress, and some that have not yet begun. Therefore, as a percentage of the total job, the activities on which progress is to be measured often do not represent the major component. So using a network as the basis for our judgements on progress is somewhat more valid than just looking at the project as a whole. The accuracy and consistency of the progress measurement will be enhanced further if we use milestones as described below.

Milestones for Progress Measurement

If our projects tend to follow a similar pattern, it is possible to establish a set of agreed-upon guidelines tied to milestones. For example, if our projects all involve small buildings, we might set up a system as follows:

Activity	Percent of project complete when activity is complete
Survey	5
Prepare plans	15
Obtain financing	20
Obtain permits	25
Grading and excavation	30
Pour foundation	35
Frame building	45
Begin plumbing and wiring	50
Roof building	60
Install siding, windows, and exterior finish	70
Install interior carpentry	80
Finish plumbing and wiring	85
Finish interior	95
Final grade and landscaping	100

A set of guidelines based on these milestones should document what is meant by each activity's completion. For example, it should be clear exactly what work has been done when the framing building step is complete, and what work is included in each succeeding activity. The allocation of percentages can then be based on our judgement of the allocation of man-hours or costs. For progress measurements on a given day, judgement would be used to interpolate between milestones or handle situations in which several incomplete activities are in progress at once.

"Remaining Duration" to Measure Progress

When progress is measured by judgement it is best if that judgement's format relates well to the way that people think. For example, when asked, "What is your % complete?", many laymen will give an optimistic answer, not only because of their naturally optimistic bias but also because of their tendency to minimize in their minds the work remaining. However, if a field superintendent or foreman is asked, "How long will it take you to finish this work (given a certain number of people, materials and equipment)?", he can usually give a pretty accurate answer. The remaining duration of the activities in progress can then be used to update the project network and to forecast the project-completion date. The % complete of each activity is calculated as the ratio of time spent to date over total duration. Overall % complete can then be calculated in the usual way by weighting the % complete of each activity by its planned man-hours.

Use of Physical Quantities to Measure Progress

Some companies find it useful to use overall physical quantities to measure progress. For example, on a job with a lot of welding, we might use cubic in. of weld installed as a progress indicator. Knowing the total cubic in. of weld metal to be laid down, and the amount done to date, we have a rough indication of progress. Similar procedures could be followed with other physical quantities, such as in.-ft. of piping installed, tons of structural steel erected, linear ft. of cable installed, etc.

Unit-price contracts, by nature, provide a good progress measurement method, as they require physical quantities of work to be measured accurately if the contractor is to be paid. Therefore, in this case, the progress-measurement system should be compatible with the schedule of unit prices, such that the quantity measured for contract administration gives us an automatic input for progress measurement.

CHAPTER SUMMARY

Effective project control requires some measurement of physical progress. An understanding of the basic principles of the earned-value system is essential even if progress measurements based on judgement are to be used. For small projects, approximate methods can be used effectively provided the guidelines are clearly defined and consistently applied. Planning networks provide a good basis for a number of shortcut progress-measurement techniques.

9
Control of Cost, Schedule, and Resources

INTERACTION OF COST, SCHEDULE, AND RESOURCES

In "Integration of Cost, Time, and Resources," in Chapter 2, the concept of integration was introduced, and the inherent dependencies of cost, time, and resources explained. Nowhere is this dependency more noticeable than in the area of project control. We can easily observe that if we seek to tighten our schedule, the costs will probably increase, as will the amount of manpower required. If we fail to provide sufficient manpower for the job, both schedule and cost are likely to suffer. Additionally, if we seek to minimize costs, we often do so by applying minimum resources to the work.

Effective project control, therefore, requires an integrated approach. Fortunately, our budget plan (the project model) provides us with the ideal basis for integrated project control. In Chapter 2 we saw that an integrated plan—our project model—tells us, for each activity:

What scope of work is performed
What material, manpower, and equipment resources are required
How much time the work is expected to take
What it will cost (to perform that scope of work, with those
    resources, in that period of time)

As the project progresses, the actual scope, the actual duration, the actual resources applied, and the actual cost will all be somewhat different from the plan. The reasons for these differences are the project trends and developments which we seek to control.

From the previous two chapters, we can define the essential elements of an effective project-control system:

A means for timely management reporting
A progress-measurement system
A basis against which to compare actual costs, times, and resources

For small projects, these essential elements may, if necessary, be
provided in a rough, approximate way. The important thing is that
each be provided.

BASIC PRINCIPLES OF FORECASTING
AND CONTROL

A definition of the proper basis of a cost or schedule forecast was
provided in "Using the Tracking Curve for Forecasting and Control"
in Chapter 7, and is worth repeating here:
  *A forecast predicts the final outcome of the project if the trends
identified to date continue to the project's completion.*
  To put it another way, we can say that the forecast represents the
final outcome of the project if nothing is done to correct the problems
or adverse trends identified to date.
  We might argue that a forecast should be like the original cost estimate and attempt to predict the "expected" outcome. This point of
view has some validity, but were we to do so we would have to assume
that some action would be taken to correct the problems at hand. That
is precisely where the problem with this approach lies: if it is assumed
that effective corrective action will be taken, chances are it never will
be. Therefore, by forecasting the results if things continue the way
they are going now, we call attention to the need for corrective action
as well as to the consequences of such action not being taken or not
being effective. If that result is unacceptable, the need for action is
clear and action is likely to be taken. Our cost/schedule forecast is the
way in which we focus the "spotlight" on the areas requiring
management attention.
  To make a forecast and identify problem areas, we need to ask some
important questions. How well we answer these questions will determine
how well we control the project. The questions that need to be asked
with regard to schedule; design and construction quality; labor,
material, and equipment resources; and costs are:

Where did we plan to be at this point?
Where are we now?
If there is a difference between our planned and actual situation,
  what is the reason?
If things continue as they are, what will be the result?
Has corrective action, applied to problems previously identified,
  been effective?

What are current problems and what can be done to correct them?

We will see specific examples of this measuring, comparing, analyzing, and forecasting process later in this chapter.

SCHEDULE FORECASTING AND CONTROL

Assessing Current Schedule Status

One way to assure that we stay "on top" of the project is to regularly review its status. It is helpful to conduct these reviews, whether quick or in depth or done alone or with others, by using a checklist. A checklist of the project-control parameters which affect the schedule is as follows:

Physical progress to date
Productivity (the relationship between physical progress and actual man-hours)
Maintenance of schedule milestones
Availability of materials
Availability of labor
Provision of overhead services (supervision, transport, etc.)
Progress and productivity trends to date

In order to accomplish schedule control we must evaluate how the project stands with respect to each of these parameters in comparison to the original budget plan. The more complete our budget plan is, the better we will be able to identify specific areas of variation and specific reasons why the variation has occurred. For example, if we have an integrated budget plan which shows the schedule for the application of field manpower, we can quickly ascertain whether any delay in progress was due to inadequate manpower by comparing planned vs. actual manpower utilization (see the computer-generated graphics in Figure 9.1). Such a comparison also indicates which crafts and which time periods were involved.

As a minimum checklist for small projects, we should ask ourselves:

Is progress as planned? If not, why not?
Are the man-hours expended as planned? If not, why not?
Are materials being shipped and received on schedule?
Are major milestones still on schedule?
Have any major problems been identified? If so, are they being handled properly?

As before, our answers to these questions point the way to the actions required for effective control.

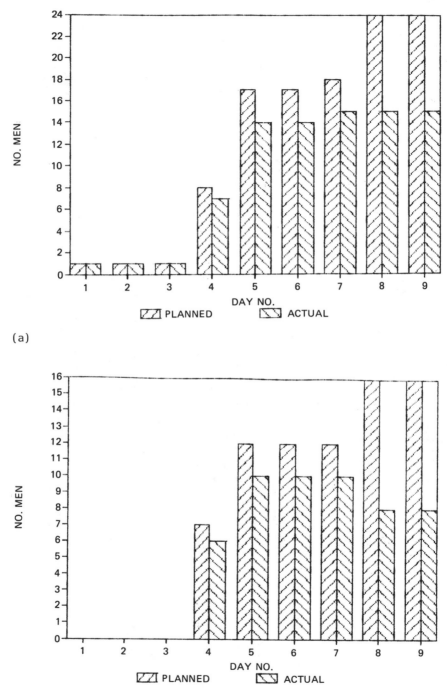

FIGURE 9.1 Manpower utilization. (a) Total. (b) Laborers.

Forecasting Schedule for Completion

Once we have completed our review of current schedule status, we
must prepare an updated forecast of the schedule for the work remaining. To do so, we must reexamine all the assumptions behind our
original schedule, and redefine the basis for the work still left to do.
It is important to remember that the schedule forecast must be consistent with the current schedule for the application of manpower and
materials resources.

There are various methods for schedule forecasting. The appropriate
method for a small project depends on the level of detail of the budget
plan, the importance of the project, and the time available for forecasting an control. Some methods are discussed below.

Updated Network and Resource Analysis

If the original budget plan and the project-control methods are computerized, it is a relatively simple matter to run the scheduling program
for the remaining scope of work. The basic steps to prepare the updated schedule are as follows:

1. *Check the network logic.* Mandatory constraints will probably
still be valid, but arbitrary constraints may well be subjected to
change. In the case in which the schedule has slipped but the completion date must be maintained (not an uncommon situation), it is often
useful to reexamine the constraints. There may well be activities
which can be done in parallel, although it may be more inefficient and
costly to do so.
2. *Review past and future milestones.* If past milestone dates have
slipped, future milestone dates are also likely to slip unless the cause
for the previous slippage has been identified and corrected.
3. *Check the delivery dates* (for materials required for the work
left to go). If delivery dates have slipped, check network float to see
if slippage can be accomodated without changing the schedule.
4. *Check the availability of manpower* (as required for the work left
to go). If manpower shortfalls are likely, revise activity durations accordingly and/or update the resource analysis to reallocate manpower
between activities.
5. *Input actual start and completion dates* for activities that have
already begun.
6. *Input actual progress* and remaining durations.

Use of Average Rate of Progress

On small projects for which an in-depth schedule analysis may not be
justified, we can simply calculate the average rate at which we have
made progress to date (e.g., 10% per week) and assume that the rate
progress will continue to the end of the job. Such forecasts should use

the typical S-shaped progress curve, and provide for a slower rate of progress towards the end of the work.

Use of Milestones

On small projects without a proper plan and schedule, milestone dates can be used to indicate the liklihood that the planned completion date will be achieved. As mentioned above, if the planned milestones to date have slipped, chances are the final date will slip also unless the causes are identified and corrected.

Use of Tracking Curves

The use of tracking curves to define current status and forecast the work left to go is discussed in Chapter 7.

FORECASTING AND CONTROL OF RESOURCES

Importance of Resource Management to Small Projects

As discussed in Chapter 4, small projects are particularly sensitive to the cost and schedule effects of resource shortfalls because of the limited number of alternative ways to make progress. If, on a large project, labor, materials or equipment are not provided as planned, it is often possible to make progress on other parts of the project where resources are available. This degree of freedom is not available on a small project, and often the lack of critical resources can stop progress altogether. Planning, tracking, forecasting, and controlling the utilization of resources is therefore of particular importance on the small project. As in other aspects of project control, the budget plan provides the basis for control.

Small projects may obtain design engineering services, materials, construction labor, and equipment either in-house or from outside contractors and suppliers. In some cases, a single project will have resources supplied from both inside and outside sources. Although it generally is someone else's job to do the purchasing or contracting, it is the project engineer's responsibility to see to it that these resources are indeed provided. This supervisory task is often discharged more effectively when a clear and specific schedule is agreed upon in advance, and then used as a basis for tracking, forecasting, and control. The resource schedule is described in Chapter 4. As a minimum, the project engineer should check that:

The quantities and types of materials, labor, and construction
equipment that are currently required are on the job
The quantities and types of materials, labor, and construction
equipment that are required for the upcoming work will be provided

Any problems relating to materials, labor, and equipment have been
identified and corrective action has been taken

Forecasting and Control of Labor

Labor falls, in general, into three categories:

Engineering (engineers, designers, and draftsmen)
Direct labor (welders, pipefitters, laborers, etc.)
Indirect labor (supervisors, inspectors, equipment operators, etc.)

As illustrated in Figure 9.1, we can use the resource schedule provided in the budget plan as a basis for tracking the actual application
of labor resources. At any point in the project, we can update the re-
source-analysis process by which we optimized the original time and
resource plan to reflect the man-hours spent to date, the physical
progress to date, the current schedule for the work remaining, and
the anticipated availability of labor for the remaining work (see Figure
9.2). It should be noted that changes to the schedule for the work
left to go will have an effect on the requirement for labor resources.
When updating the schedule for the provision of labor we need to
consider the following factors:

Physical progress to date, which indicates how much work remains
Labor productivity to date, which indicates the manpower required
to perform the remaining work
Contractor performance in providing labor, which indicates whether
we need to seek alternate sources of labor
Schedule of work to go, which indicates when the labor will be
required

Control of labor can be thought of as tracking and forecasting the
quantity and quality (i.e., productivity) of the manpower applied to the
job. On a small project detailed consideration of these factors may not
be practical, but they should be considered even if only by using
judgement.

Forecasting and Control of
Construction Equipment

On a small project it is generally not necessary to schedule and control
all types of construction equipment. In general, such equipment as
trucks, welding machines, small tools, etc., are readily available. (Projects in remote locations are, of course, exceptions.) The type of construction equipment which requires some planning and control has one
or more of the following characteristics:

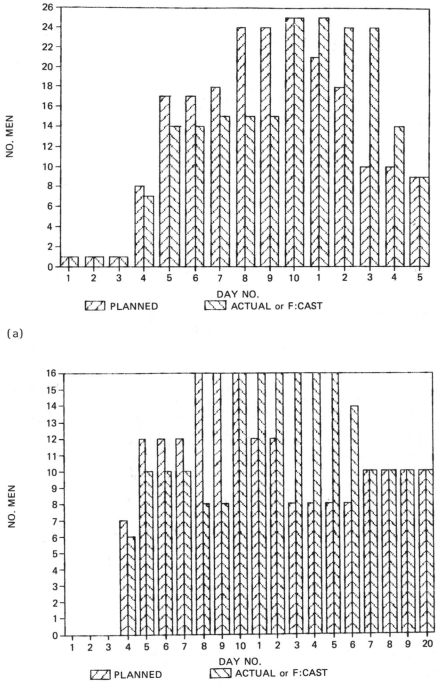

(a)

(b)

FIGURE 9.2 Manpower utilization. (a) Total. (b) Laborers.

Is not readily available
Is critical to the schedule
Requires some time to mobilize
Interferes with other operations

For example, our small project may involve lifting and placing a compressor. This activity, on the critical path, requires a large crane. The crane has a great cost and schedule impact on the project since, if the crane is not available when needed, the entire project will be delayed, and, if the crane stands idle even briefly, the costs will overrun. The crane's operation must also be carefully planned to avoid interference with existing facilities and operations. If the schedule should change (due, for example, to a delay in shipment of the compressor), changes to the arrangements for the crane must also be made quickly if charges for idle time are to be avoided. This crane is the type of construction equipment which requires control.

The techniques for forecasting and control of equipment resources are much the same as those described above for labor. As the schedule is updated, the requirements for and availability of equipment should be reviewed, and the equipment schedule revised accordingly.

Control of Materials

The various operations involving the provision of materials to the job are generally handled by the in-house procurement function, or by the contractor. These operations include purchasing, expediting, inspection, traffic, and warehousing. The project engineer's role is to coordinate and supervise these activities. In so doing, he often finds himself at a disadvantage. Unlike labor and equipment resources, materials are an "all or nothing" proposition: they are either there or they are not. Until they are actually on the site, the project engineer must rely on information from various sources to figure out what is going on, and what to do so he can assure that the material will be there when required. Often that information is a "pink copy" carbon of a purchase order, memo, or receipt notice: hardly the kind of timely information that is required. What is required is a simple but effective method for capturing and presenting information about the status of materials so the project engineer can take action if required.

Planning for Materials

Materials can be divided into three categories according to their importance to the schedule:

Long-lead materials
Critical materials (those materials are on the critical path and may or
   may not also be long-lead items)
Materials with float

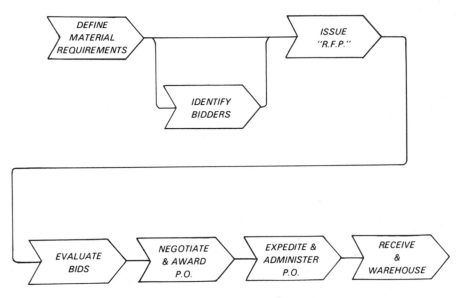

FIGURE 9.3 Material procurement plan.

The first step is to identify the materials in each category, which can be easily done using the schedule-analysis principles described in Chapters 3 and 4. The next step is to recognize that material procurement has a schedule of its own—a subnetwork to the project schedule—and that all materials follow this plan, as shown on Figure 9.3. From Figure 9.3 we also note that material purchased for the project requires more time than material which can be obtained from the company warehouse. The project engineer's job then becomes one of tracking where each material item is on its procurement schedule, starting with the critical items, then the noncritical long-lead items, then the items with float. It might also be helpful to prepare a separate schedule showing the key milestone dates for placing the purchase orders and for delivery of materials, and use this schedule for discussions with the procurement function, as well as to track progress in procurement.

If project control is computerized, it is possible to set up a database, possibly tied to the network, which presents an up-to-date summary of material status. An example of such a database is shown in Figure 9.4 The material data can be linked to the network by specifying the required material as a resource requirement for the appropriate activity,

COST CODE DESCRIPT.	P.O.NUMBER	ITEM NO.	ITEM DESCRIPTION	QTY UNIT	QTY	VENDOR	DELIVERY DATE	DATE REQUIRED	VAR. DAYS	CURRENT PRICE	BUDGET $	VARIANCE $
===	===	===	===	===	===	===	===	===	===	===	===	===
211 HEAT EXCH.	A23-444-1	1	E-101	EA	1	JONES & GREEN	JULY 23	JULY 25	-2	$23,500	$25,500	($2,000)
		2	E-102	EA	1							
213 PIPING	A23-445-2	1	PIPE 10"	FT	160	T.H.E. TUBES	JULY 15	JULY10	5	$34,500	$30,700	$3,800
		2	PIPE 8"	FT	220							
		3	PIPE 6"	FT	320							
		4	ELLS 6"	EA	12							
		5	TEES 6"	EA	10							
	A23-448-4	1	GATE VALVE 10"	EA	2	STOPFLO	JULY 8	JULY 8	0	$17,500	$19,500	($2,000)
		2	BALL VALVE 8"	EA	1							
224 PUMPS	A23-457-4	1	P-101	EA	2	GPM MANUFACT.	JUNE 30	JULY 15	-16	$12,500	$10,200	$2,300

FIGURE 9.4 Small-project material-status database.

and the computer system can then link the network activity to the
procurement information in the database. The key to the success of
such a system is to capture data from existing documents, and to avoid
creating new procedures.

Controlling Materials

For the small project, for which we are not likely to have a sophisticated material-control system, we can use a simple checklist to help
control the materials function:

Have design specifications and purchasing requisitions been com-
   pleted on schedule?
Have purchase orders been placed on schedule?
Has everything that has been requisitioned to date been ordered?
Have the correct items and quantities been ordered?
Are the anticipated delivery times in accordance with the schedule?
Are immediately-required materials available?
Does the schedule for purchase and delivery of materials reflect the
   latest update to the project schedule?
Do the current material requisitions reflect the latest design changes?
Have problems related to materials been identified and corrective
   action taken?

As the materials status is periodically updated, the overall project
schedule must, of course, be revised accordingly.
   On of the interesting uses of float in a schedule is in assessing the
impact of anticipated delays in the delivery of materials. On a project
without a network plan and schedule, delays in materials are often
interpreted as a problem requiring immediate corrective action. Time
and money are often spent devising ways to improve the delivery times.
In some cases, however, a network-based schedule will show that the
material delivery slippage can be accomodated within the existing float,
and that corrective action is not necessary. Therefore, network time
analysis helps not only by focussing attention on those items which
are critical, but by avoiding panic situations over items which are not.

Allocating Fixed Resources to Multiple Projects

In the small-project environment, in which we often have many projects
underway simultaneously, the main problem involving resources if often
that of allocation between projects. For example, the company may
have a Drawing Office which does the design and drafting work for each
of the small projects, and the problem is to allocate the limited resources
of that office to the various projects. Or, the project-engineering department itself may consist of a number of project engineers, each of
whom handles several projects, and the problem for the department
manager becomes, "To whom should I assign this new product?" Often

the company has a pool of labor forces and construction equipment, and these resources have to be allocated among the various projects. This problem is discussed in "Allocating Resources to Multiple Projects," in Chapter 4.

Some companies find it helpful to use a priority system for small projects. This often works well, but the problem with such systems is often that lower-priority jobs never done until they slip so badly that their priority is upgraded. This undermines the priority system and tends to make it based more on office politics than on the actual priority of the projects.

The concept of the network heirarchy was introduced in Chapters 2 and 3. This technique, in which we use "hammocks" to summarize several activities and to create a higher-level network, allows us to create a "project" which is, in effect, the work required to do all the projects currently defined. Such a "Level 1" network is illustrated in Figure 4.9. Since all the projects shown as activities cannot be executed simultaneously, we can schedule them according to the availability of resources. Using resource-limited scheduling, we can prepare a resource allocation schedule (see Figure 4.9). As new projects are added, current projects slip, and existing projects are completed, the Level 1 schedule is changed accordingly. This procedure requires a computer system, but, assuming one is available, provides a most effective way to handle allocation of fixed resources to multiple projects.

COST FORECASTING AND CONTROL

What is "cost control?" To some, it is the recording and analysis of cost data such as timesheets, purchase orders, and invoices. Such data is historical, and, as such, it makes an essential contribution to cost control. However, no one can control anything solely by concentrating on what has already happened. We will refer to cost control in the context of the things we do to affect the current and future activities of the project, and, hence, the final outcome.

The final cost of a project, large or small, is simply the result of the many things that were done to accomplish the actual scope of work with the resources that were used over the project's duration. Thus the cost is a reflection of what was done and how well it was done, given prevailing economic conditions. Therefore, it might be said that *there is no such thing as cost control!* We cannot control costs, but we can control the things that happen on a project to determine its cost. We can assure that the cost impact of the various decisions that are made is recognized as part of the decision-making process. We can assure that unnecessary costs are not incurred due to poor productivity, schedule slippage, inefficient allocation of labor resources, unnecessary design or field changes, and so forth. We can, finally, assure that we have a well thought out and documented planning and cost basis for use in tracking and controlling the project, as well as the appropriate methods, systems, and procedures. In fact, all the material presented so far has to do with cost control.

Cost control, then, consists primarily of assuring that the project is managed with full recognition of the cost impact of everything that is done, and everything that happens. In this section we will review some basic principles and aspects of cost control.

Analyzing Variations to the Estimate Basis

In "Establishing the Estimate Basis," in chapter 5 the importance of the estimate basis was discussed. The estimate basis can be thought of as a "tripod," the three "legs" of which are the design basis, the planning basis, and the cost basis (see Figure 5.2). The basis, there-fore, describes what is to be built, how it is to be built, and the pricing levels that are anticipated. All variations between the actual costs and the estimate can be explained in terms of variations to this estimate basis. In "Reconciling Forecasts and Estimates" (Chapter 5), the use of reconciliations was discussed as a control device in which the variations to the estimate basis were identified and costs attached to them to explain the differences between actual, estimated, and forecast costs.

One aspect of cost control, then, consists of periodic reviews of the project as it progresses, and the comparison of these reviews with the estimate basis. We might use a checklist for such reviews, as shown below. As we review each item we should ask: "How does the way the project is actually progressing differ from the assumptions made in the estimate basis? If there is a difference, what is the cost effect likely to be?"

The design basis

Specification of overall performance (capacity, throughput, production rate, feedstock, etc.)
Overall scope of work
Type and extent of design changes
Type and extent of field and startup changes
How late in the project design changes are made
Man-hours spent in design work
Extraordinary technical problems identified
Timeliness and effectiveness of design reviews

The planning basis

Schedule duration
Milestone dates

Contracting plan
Purchasing plan
Activity constraints
Use of shift work and overtime
Manpower density
Progress relative to reference schedule
Labor productivity
Engineering productivity
Availability of materials
Availability of labor
Availability of construction equipment
Availability of drawings

The cost basis

Prevailing escalation rates
Average hourly costs for engineering
Average hourly costs for construction labor
Average unit costs for materials (e.g., dollars per ton of piping)
Quantities of materials
Number and cost of changes
Contingency required
Indirect costs as a percentage of direct costs
Bulk materials as a percentage of equipment costs
Cost to achieve 1% physical progress

Tracking curves, discussed in Chapter 7, are a useful way to monitor trends and identify variances. Some additional sample tracking curves are shown in Figure 9.5 These computer-generated curves show the same information presented in x-y and histogram formats. Extrapolations from the tracking curves can be used for cost forecasting.

Cost Forecasting

The same rules of forecasting that were described in "Basic Principles of Forecasting and Control" earlier in this chapter also apply to costs. A cost forecast should reflect the expected final cost if things continue as they have been, and nothing is done to correct negative trends. If we are using an integrated cost-schedule-resources project model, the "current model" represents design change approved to date, the "forecast model" represents what is expected to happen, while the "static model" represents the budget plan and the estimate basis.

In cost forecasting, it is important to distinguish between factors affecting the project which are controllable and those which cannot be controlled or corrected. These might be called "external" or "internal" factors, depending on whether they impact the project from an external

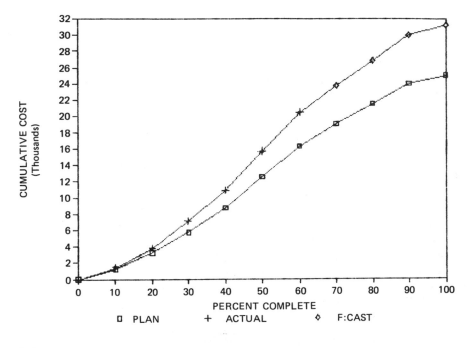

(a)

FIGURE 9.5 Cost tracking and forecasting. (a) Cumulative cost vs.
percent complete. (b) Period cost vs. percent complete.

cause (and therefore are not subject to the control of the project engineer), or whether they are internal to the project. For example,
escalation rates cannot be controlled by the project engineer, although
he may be able to take some steps to minimize their impact. Design
changes, on the other hand, can be controlled. Labor productivity
may or may not be controllable, depending on whether it is determined
primarily by market conditions (and is therefore external), site conditions (internal) or contractor performance (internal). The net cost
impact of the external, uncontrollable factors establishes a minimum
level for the cost forecast of which the project engineer should be
aware.

In general, cost forecasts are made by calculating the value of work
done, and adding the forecast cost of the work remaining. A forecast
can also be made by updating the estimate basis for productivity, wage
rates, etc., and recalculating the estimate. A key point about cost
forecasts is that they are a matter of judgement, and the forecaster
should feel free to make the forecast according to his best analysis and
judgement.

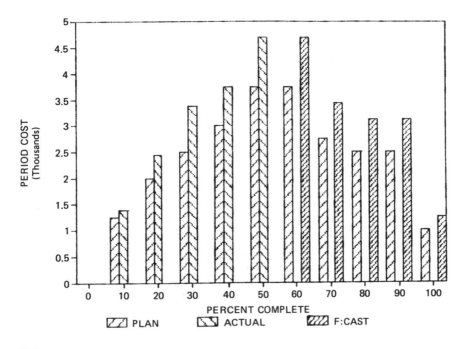

(b)

FIGURE 9.5 (Continued)

Control of Changes

One of the most frequent causes of cost overruns is changes. This is, in fact, most unfortunate because changes are generally an internal, controllable factor. Therefore, one of the most important things that can be done to effect cost control is to establish proper control over changes.

Definition of a "Change"

What is a "change?" Many spirited discussions have taken place over just that question, particularly when contract terms cause large sums of money to ride on the answer. We can define a change as follows:

A "change" is a specific work assignment which would not ordinarily be assumed to be required to complete the original scope of work. A change may also be an instruction to perform a specific work assignment in a different way from the way which was previously defined.

Changes are normally made to improve the performance, operability, maintainability, safety, or cost of the facilities.

Note that this definition excludes "design development" work from the category of changes. As design work progresses and the design is defined in increasing detail, alternative approaches to various design problems are studied, and a number of revisions to the work already done normally occur. This is normal process, and these minor revisions are not generally considered changes, even if they are initiated by the client in the course of his normal reviews. If, however, an instruction is given to do the work in a different way than might normally be assumed, or to do an item of work not normally required to complete the original scope of work, then that instruction requires a change to be approved.

Now that we have defined the basic concept of a change, we can define the different types of change (see also "Definition of Project-Control Terminology," in Chapter 7).

Design change: A change made during the design phase of the
   project which modifies work already done or adds work not normally assumed to be required to complete the original scope of
   work. A design change is initiated by the design function.
Scope change: An increase or decrease to the original scope of
   work. This generally means a change to the overall specification
   of the project (e.g., capacity, throughout, facilities installed,
   acres of land, etc.) or to a contractually-defined scope.
Field change: A change initiated in the field to facilitate construction
Startup change: A change made to facilitate or simplify the startup
   of the facilities. Also, a change made during the startup phase
   of the project.

Control of Changes

Most engineers have a tendency to change things; always, of course, to make them better. Changes are inevitable on a project and they are often initiated faster than we can keep track of them. Unfortunately, once changes get out of control, our cost-control work becomes ineffective because we can no longer analyze variances since we cannot distinguish those variances caused by changes from those caused by other factors.

Changes are often difficult to estimate as their cost impact tends to have a "ripple effect." A change which seems to affect one department may have a larger effect than anticipated. For example, suppose the hp rating of a pump is increased. In addition to the pump costing more, its driver will cost more too. The pump and driver will weigh more, possibly resulting in increased weight and cost of structural steel and foundation. The piping from the higher-power pump will

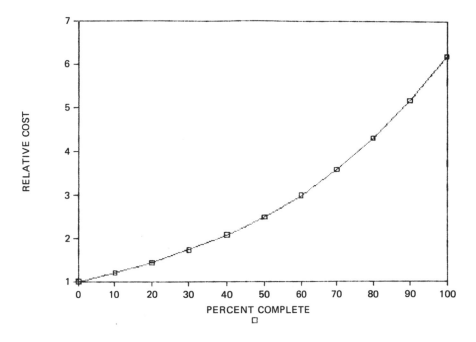

FIGURE 9.6 Cost of a change vs. time.

likely be of a higher flange rating, resulting again in more weight and cost for the piping and its supporting steel and foundations. Similarly, the electrical cables to the pump will be larger and heavier, as will the switchgear and other electrical equipment. Additional man-hours will be required in the field to install this heavier equipment and materials, and additional man-hours required to revise the design. As a result, the engineer who suggested a change to the higher-hp pump, which might cost $2000 more, is often shocked to find that the net cost impact is $20,000 when all the costs are accounted for.

The timing of a change also has an effect on its cost. As seen in Figure 9.6, a change made early in the design phase, when everything is on paper and is still preliminary, will have the minimum cost. As the design progresses, more drawings in more departments must be changed to accomodate the same design change. Finally, after construction has begun, the change might involve removal or modification of construction work already done, and that involves the highest cost.

The essential elements of a change control system are:

The ability to identify changes as they occur
The ability to prepare quick estimates of the cost impact of each
   change

[THOUSANDS]

COST CATEGORY	BUDGET ESTIMATE	APPROVED CHANGES	CURRENT ESTIMATE	FORECAST FIN. COST	VARIANCE $	%
DIRECT COSTS						
MATERIALS	175	7	182	165	-17	-9.3
LABOR	125	8	133	155	22	16.5
SUBTOTAL	300	15	315	320	5	1.6
INDIRECT COSTS						
ENG'G	145	4	149	155	6	4.0
FIELD O'HDS	75	3	78	85	7	9.0
SUPERVISION	55	3	58	55	-3	-5.0
SUBTOTAL	275	10	285	295	10	3.5
OTHER	45		45	50	5	11.1
BASE EST.	620	25	645	665	20	3.1
PENDING CHNG				12	12	
CONTINGENCY	93	-30	63	60	-3	
TOTAL	713	-5	708	737	29	4.1

FIGURE 9.7 Typical cost report format.

The requirement that a manager or project engineer with responsibility for the budget approve changes only when the cost impact
is known.

To create such a system, we need formal procedures and a quick
estimating method. It is also helpful to define the various stages of a
change, so that they can be tracked through each stage.

Potential change: A change that is still in the idea stage, but which
is likely to be initiated.
Pending change: A change for which approval is sought but has not
yet been obtained. A change is pending during the time that the
design details are worked out and the estimate prepared.
Approved change: A change for which the design and estimate have
been approved. Approved changes are usually included in the
"current control estimate," and are offset by reductions in
contingency.

Cost Reporting

In general, cost reports should have the basic elements provided in the example shown in Figure 9.7, and described below:

Cost category: This groups costs into categories for control.
Budget estimate: The budget for the project (the "static" model).
Approved changes: Shows the total approved changes to date.
Current control estimate: This is the budget estimate plus approved changes.
Current cost forecast: This is the value of work done plus cost of work to go.
Variance: This is the current forecast—current control estimate.

Note that the current cost forecast includes provision for pending changes as well as for contingency to cover the uncertainties surrounding the remaining work. Attached to the cost report might be a list of approved, pending, and potential changes, tracking curves as appropriate, a written analysis of cost trends, and a reconciliation to the previous forecast as well as to the budget.

QUALITY CONTROL

Although quality control is a highly complex subject, the details of which are outside the scope of this book, there are a number of basic principles which are relevant to the project engineer. It is logical to begin our discussion of quality by defining our terms, something that is often difficult in this subject. For the purposes of this book, we will use the following definitions (as shown in Chapter 3).

Quality: The fitness for the intended purpose. The better something fits the purpose for which it was intended, the higher quality it is. In most cases, the intended purpose of an engineered facility is to produce a profit for the company. Therefore, the item of highest quality will not always be that which is most expensive as it will be that which represents the optimum combination of price and performance.

Quality assurance: This is the standards we use to specify quality and the program of design reviews, procedures, and inspections we intend to use to assure adherence to those standards.

Quality control: This is the activities which we conduct to carry out the quality assurance plan.

Therefore, in the small-project environment, we set up a quality-assurance program for all small projects, and conduct quality control on each one. The basic activities involved in quality control on a small project are:

Design reviews in which the operations and maintenance personnel who will operate the facilities make comments and suggest changes to the design. The purpose of such reviews is to incorporate the design input of the user at an early stage in the design process to minimize the cost of such changes, as well as to assure adherance to standards and specifications.

Inspection and testing in which shop- and field-fabricated equipment and facilities are tested and checked.

Certification in which approval is obtained from a certifying authority (e.g., nuclear and marine projects)

Change control as described above.

We can use our project model for quality control by using it to identify, plan, and schedule the specific quality-control activities required. We can then monitor actual performance to assure that the required inspection, testing, and other quality-control activities are performed. If our system is computerized, we can set up a simple database to keep track of the relevant documents to assure that the proper certificates can be produced at any time.

Given the broad definition of quality we are using, we can see that the capital and operating cost of the project have a great deal to do with its profitability and hence with its quality. Of the three legs of the estimate basis (design, planning, and cost), design has the greatest impact on project cost, and it is completely controllable internally. Therefore, good control of changes and close monitoring of the cost implications of design trends and decisions is an effective form of quality control.

CHAPTER SUMMARY

In this chapter we discussed the various aspects of project control. We saw how the integrated-project model provides an excellent basis for control and how important resources—labor, materials, and equipment—can be monitored and controlled. Much of the effectiveness of the project-control function relies on the planning, scheduling, and resourcing work done prior to the start of the project. This work makes it relatively simple to track performance, identify problem areas, and make forecasts.

Part IV
Dealing with Uncertainty

10
Quantifying Estimate Risk, Uncertainty, and Contingency

CONTINGENCY: AN IMPORTANT PART OF EVERY ESTIMATE

All those who prepare cost estimates share the common dread of being called upon to answer the question, "How much contingency should I add?" In actual practice, the answer to that question really depends on the answers to two more specific questions:

1. How much contingency is really needed?
2. How much contingency can we get management to accept?

The whole subject of contingency deals with uncertainty. It is therefore appropriate that most project and cost engineers experience a lot of uncertainty when dealing with it. We are usually uncertain about (among other things):

What contingency is really for
How to define it
How it relates to risk
How it relates to estimate accuracy
How to estimate it
How to control it
How to explain it
How to justify it
Where to put it
What to call it

Trying to explain contingency to management is a lot like a visit to the dentist: the best that can happen is that you come out the same

way you went in—all other possibilities are increasingly painful. Even
the word "contingency" has negative emotional connotations. It implies
risk and uncertainty which cause fear.

Like it or not, contingency, risk, and estimate uncertainties are all
most relevant today. Capital projects today are taking place in an
environment of greatly increased financial risk, and of increased
emphasis on budget levels, cost control, and effective use of
management tools. Projects today, as compared to those in the past,
are often:

Marginal, posing more risk to profitability
Longer term, with profit further in the future
Based on new technology, with associated risks
Subject to larger market fluctuations

Therefore even for small projects, there is a need to develop a clear
and practical approach to handling contingency.

THE UNCERTAIN NATURE OF ESTIMATES

Although most project engineers and managers deal with estimates every
day, it is worthwhile to reiterate just what an estimate really is. According to Webster's, to estimate is to "calculate approximately the
worth, size, or cost." So an estimate is, by definition, an approximation; a prediction of what is going to happen. Therefore, *all estimates,
by definition, contain a significant amount of uncertainty*. In spite of
this indisputable fact, most project engineers and estimators tend to
"sell" their estimating work by making the estimate look as credible
as possible, often by downplaying (or even covering up) uncertainties.
This is a natural tendency: it's hard to find a "how to succeed" book
that recommends telling the boss how uncertain we are about the work
we have just done and are asking him to use as the basis for a major
decision.

The reason for this paradox is found in the way estimates are used.
The basic purpose of estimates is to aid in decision making. Company
management uses estimates for decisions such as:

Selection of projects for further study
Selection of the preferred design approach
Economic analysis
Budget approval and commitment of funds for engineering, procure-
    ment, and construction
Selection of contractors and suppliers
Approval of major changes

Project managers and cost engineers also use estimates for the very
important purpose of cost control.

All these estimate users require accuracy in their estimates. They
want to make feel confident that they have good estimates on which to
base their difficult decisions. Therefore, estimate uncertainty, accuracy, risks of overrun, and contingency are subjects which are apt
to be avoided in conversation. The project engineer or estimator
doesn't want to talk about it, and the manager or estimate user doesn't
want to hear about it. This would be a happy situation were it not for
the unfortunate result that *many important investment decisions are
made without proper recognition of the financial risks involved.*

It is, therefore, important that project engineers and managers be
able to quantify, control, and communicate risk, uncertainty, and contingency. Whether project engineers or managers like it or not,
estimating means dealing with uncertainty and decisions based on
estimates mean dealing with risk.

A DEFINITION OF CONTINGENCY

Before we can measure something, we have to be able to define it.
Unfortunately, everyone knows what contingency is, but everyone
thinks it is something different. For example, there is one school of
thought (illustrated by Figure 10.1) which maintains that contingency
is a source of funds, there if you need it, which shouldn't be used
under ordinary circumstances. The problem with this view is that the
contingency funds have a way of getting spent before the project is
over.

Another view of contingency (illustrated in Figure 10.2), suggests
that contingency should be rationed out to project managers or contractors when they show that they have done their jobs, and the requirement for extra funds is not their fault. This approach is intended
to make them work for any releases of contingency. The problem here
is apt to be that the contingency is needed anyway, but that the need
for it may be concealed.

Another popular approach, illustrated by Figure 10.3, maintains that,
"contingency is not required on our projects, and we don't allow it."
Naturally, these projects do get by without contingency funds, although
they often have what amounts to contingency funds given a different
name.

At the other end of the spectrum, we can identify the approach in
which contingency funds are expected to be spent, and are spent,
freely. This approach, illustrated by Figure 10.4, naturally tends to
make cost control a problem.

As a result of these diverse views, which sometimes co-exist in the
same organization, many project engineers adopt defensive strategy
shown in Figure 10.5. Just like the family dog, which buries bones

Contingency: Funds which are not
to be used under
ordinary circumstances

FIGURE 10.1 The "piggy bank" view of contingency. Contingency is,
in this view, funds which are not to be used under ordinary
circumstances.

where only he can find them to assure that they are there when needed,
we "bury" contingency funds in our estimates and cost forecasts.

So it is an important first step just to define contingency. The following definition of contingency (presented in Chapter 5) has proven,
in the author's experience, to be clear, useful, and generally acceptable:

Contingency:
Something you
can have if you've
been good

FIGURE 10.2 The "cookie jar" view of contingency. Here, contingency
is something you can have if you've been good.

FIGURE 10.3 The "ostrich" view of contingency. Here, the existence of contingency is completely denied.

Contingency is a provision for those variations to the estimate basis which are likely to occur but which cannot be specifically identified at the time the estimate is prepared.

There are three key elements to this definition:

1. *The basis of an estimate is the design, the execution plan, and the pricing levels* that are defined or assumed at the time the estimate is done. The actual costs will vary from the estimate to the exact extent that variations are experienced in the basis. Therefore, we can explain all uses of contingency funds in terms of variations to the estimate basis.

2. *Contingency covers those variations which are likely.* In other words, low probability occurrences, such as *forces majeures*, are not covered by contingency, even if they could have a major cost impact. Since the project budget should include contingency, it should only

Contingency : There to be spent

FIGURE 10.4 The "blank check" view of contingency. Here, contingency is seen as funds that are there to be spent.

**Estimators "Bury" contingency
where only they can find it**

FIGURE 10.5 Estimating contingency: the "backyard bone-burying" approach. Estimators "bury" contingency where only they can find it.

provide for those added costs which have a reasonable probability of occurring.

3. *Contingency covers those variations which cannot be specifically identified.* In other words, if we can identify a likely variation, such as a major design change, it belongs in the base estimate, not contingency.

Some examples of likely variations that, by this definition, would be covered by contingency are:

Design changes (excluding scope changes)
Variations in the way the project is managed
Estimating variations and errors

Equally important are the potential variations that are not covered, under this definition, by contingency. These are the low probability, high cost-impact risks that are not likely to occur but, if they do, can cause enormous disruption. Examples of such risks are:

Scope changes
Major changes in schedule milestone dates
Unprecedented fluctuations in exchange rates, inflation rates, etc.

Although it would clearly be inappropriate for the project manager's budget to contain funds for such risks, it is entirely possible that top management might wish to know what the financial impact of such risks would be, and to have some funds in reserve for that purpose. Other investors in the project, or its financiers, might want to know, "How far could I possibly be sticking my neck out?" To answer this question, we define "manager's reserve": the extra funds required to produce a very high confidence that the project will not overrun.

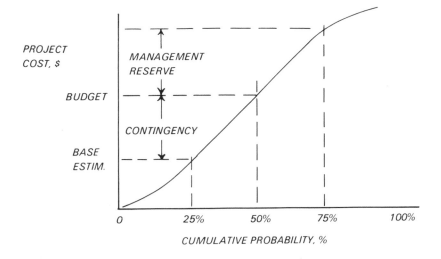

FIGURE 10.6 Project cost vs. probability.

To illustrate these concepts, we might plot cumulative probability against project cost (see Figure 10.6). A cumulative probability of 50% means that there is a 50% probability that the final cost will be equal to or less than the cost shown on the curve at that point. If the project manager's budget is to be at the 50/50 point (i.e., for him to have an equal chance of overrunning or underrunning) the contingency will be the difference between the cost at the 50% probability point and the base estimate excluding contingency. Management reserve will be the difference between the project manager's budget and the cost at the point of desired confidence.

WHY CONTINGENCY IS NECESSARY

The above definition raises some interesting questions: "Why should contingency be required?" "Why aren't the base estimates at the 50/50 point?" "Why are the variations likely to produce a net increase in the project cost?" Or, to put it another way, "Why does Murphy's Law always seem to be proven right on my projects?" To answer these questions, we need to use the statistical concept of "skewness." If we were to plot the frequency with which we expect to experience each possible value of project cost, we would describe a "frequency distribution curve," as shown in Figure 10.7. We would find that there is one value which will occur more often than any other value. This "most likely" value is the highest point on the frequency distribution

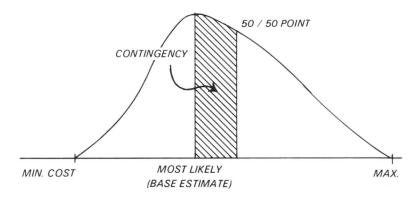

FIGURE 10.7 Skewness of project costs.

curve, and is usually our base estimate. Thus, if we look at a component of the estimate (such as material costs), at an estimating variable
(such as labor cost per hour), or at the total project cost, we generally
find that we have, using the "most likely" value, estimated the cost
which is most likely to be correct if the basis of the estimate remains
unchanged. Unfortunately, variations to the estimate basis are certain
to occur.

If we now consider the range of possible variations, we note that the
best-case scenario gives us a minimum cost somewhat below the base
estimate. There is, of course, a minimum cost level below which we
cannot go. However, when we consider the maximum possible cost
under the worst-case scenario, we find that there is apt to be almost
no limit to how bad things can get and how high the cost can go. For
example, although it is impossible for the project to be completed at
100% below budget, there are many projects which have ended up
several hundred or even several thousand percent over budget.

So skewness is an inherent part of project-frequency distributions.
Most project data is skewed to the right, meaning that there is a
greater probability that our base estimate will overrun than that it will
underrun. This is shown in Figure 10.7, in which we see that the 50%
probability point is to the right of the base estimate. The difference
between the most likely estimate and the 50/50 estimate is contingency.

THE ESTIMATOR'S DILEMMA

Having formulated an acceptable definition of contingency, the estimator
is now faced with the task of quantifying contingency: that is, of
defining how much it should be for a given estimate. However, if he
accepts the definition provided above, and begins to define a value for

contingency, the estimator finds he is faced with a dilemma: How can
one define that which, by definition, is undefined? In other words,
if contingency is for those variations which cannot defined, how are we
supposed to define them sufficiently so that we can put a value on
contingency?

This dilemma is confusing enough, but if we try to work with it, it
gets even worse. It is apparent that some form of statistical analysis
is necessary if contingency, risk, uncertainty, and estimate accuracy
are to be quantified. That itself is a problem: for some reason, very
few people are comfortable with applied statistical methods. For
example, these people say that a statistician will insist that a man with
one foot encased in a block of ice and the other planted on a bed of
burning coals is, on the average, comfortable! Most company procedures, in addition, require that a cost estimate or budget have one
value. Ranges of values are not permitted or appreciated.

On top of all this, it is also important to know that *the perfect method
for analyzing risk, uncertainty, estimate accuracy and contingency
does not exist!* Therefore, whatever is done to improve the way things
are done will still be vulnerable to valid criticism. So what is the
estimator or project engineer to do?

DEFINITION OF RELEVANT TERMS

Improving the way we define, communicate, and manage contingency
requires the same type of program we would apply to any other area
of management. Key elements are as follows:

Obtain top management support and endorsement
Define and agree on all relevant terminology
Develop and use practical methods for quantification of risk,
    uncertainty, accuracy, and contingency
Develop and use practical methods for the control of risk and the
    administration of contingency funds.
Monitor effectiveness of methods, and refine as necessary

Contingency Terms and Concepts

Let us start with the definition of terms. Like the previously-offered
definition of contingency, the following definitions have been found
by the author to be practical and acceptable.

Risk: The probability that a certain undesirable outcome will occur.
Example: There is a 10% risk that the final cost will be more than 30%
above the budget.

Uncertainty: The range of values within which the actual value is
expected to fall. Example: The uncertainty in our estimate of wage
rate is such that we expect it to fall between $10 and $40 per hour.

Variation: The difference between the design, planning, and cost basis of the estimate, and the actual design, execution, and cost of the project. All variations in cost can be explained by variations to the estimate basis. (Example: Design changes resulted in a 5% cost increase. Increased use of subcontract labor resulted in a 9% cost saving. Escalation was less than anticipated, resulting in a cost savings of 6%).

Estimate accuracy: The confidence limits within which there is a specific probability that the actual cost will fall (see Figure 10.8). (Example: The accuracy of this estimate is such that there is a 90% probability that the final cost will be within plus or minus 10% of the estimate.)

Note that this definition of estimate accuracy means that it is not adequate to describe accuracy as, "plus or minus 10%." One must always specify the probability that the actual cost will fall within the confidence limit.

Statistical Terms and Concepts

Since we will be using some simple statistical concepts in our work with contingency, it is also appropriate to review some basic statistical concepts and terminology.

Frequency distribution curve (or probability density function): This type of plot, shown in Figure 10.9, describes the frequency with which a given value will be experienced. The higher the curve at a given point, the more often we can expect that value to occur. The well-known "bell curve" is an example of a symmetrical frequency distribution.

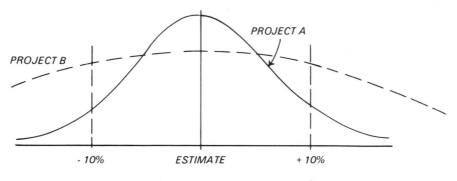

PROJECT A: 90% PROBABILITY OF BEING WITHIN ± 10%

PROJECT B: 60% PROBABILITY OF BEING WITHIN ± 10%

FIGURE 10.8 Defining estimate accuracy: the confidence limits within which there is a specified probability that the actual cost will fall.

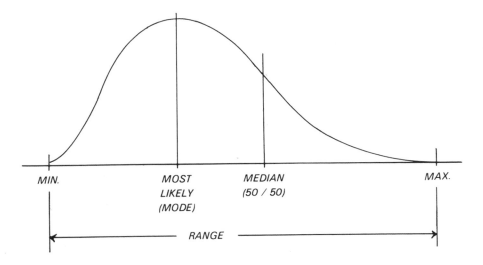

FIGURE 10.9 The frequency distribution curve.

Since the likelihood that a specific, discrete value of cost will actually be experienced is quite small, it is more useful to think in terms of the probability that the actual cost will be equal to or less than the given value. That probability is indicated by the area under the curve.

The frequency distribution curve can be described by several, very useful parameters (see Figure 10.9):

The *most likely* value is at the peak of the curve, and is the single value that will be experienced most often. It is sometimes also called the mode.

The *median* value is the midpoint of the distribution of all possible values. It is the "50/50 point," in that half of all possible values are above it, and half below.

The *mean* value is the average of all data points in the distribution.

The *maximum value* is that value for which there is a zero probability that any actual value will exceed it. It usually corresponds to the worst-case scenario.

The *minimum value* is that value for which there is a zero probability that any actual value will be less than it. It usually corresponds to the best-case scenario.

The *range* is the difference between the maximum and minimum values.

The *standard deviation* describes the "*closeness of fit*" of the distribution. If the standard deviation is large, the points are widely scattered, and the range is large. If the standard deviation is small, the points are clustered about the middle, and the range is small.

From a project engineer's viewpoint, the most important parameters are the most likely, median, and standard deviation.

QUANTIFYING ESTIMATE RISK, UNCERTAINTY, AND CONTINGENCY

Basic Approach

There are only two ways to define the probability function that defines estimate accuracy, risk of overrun, and contingency requirements. The first way might be described as "looking back," that is, by the use of historical data from many past projects. The second way might be described as "looking forward" that is, by probabilistic analysis of all possible outcomes of the project at hand. These two approaches are described as follows.

Analysis of Historical Data

This approach is based on the assumption that the project at hand is essentially similar to the many past projects represented by the data. To state it a different way, it assumes that the past is the best predictor of the future. This assumption is likely to be valid when the project at hand is considered typical. In such cases, past data from actual projects is the best predictor of what can be expected from future projects.

To prepare the cost/probability curve using historical data, the data must be adjusted to a consistent basis, and then analyzed using standard calculations for mode, median, and standard deviation. When these parameters are known, the distribution is defined and one can estimate the probability of overrun. The distribution variable which is selected can be total cost, unit cost (e.g., $ per unit of production), or contingency amount (calculated as the difference between actual and estimated costs).

Successful application of historical data analysis requires that the project be similar to past projects, that the data used is reliable and appropriate as a predictor of future performance, and that no special risks need to be reflected in the analysis.

Probabilistic Cost Analysis

In probabilistic cost analysis, the project at hand is analyzed by the process of looking into the future and examining what possible outcomes exist. In other words, we synthesize the cost vs. probability function from the specific risks and uncertainties that are associated with the project. This approach is preferred in those situations in which, (1) the project at hand is not similar to many past projects, (2) in which the risks it will be facing are specific and atypical, and/or (3) in which the available data is not adequate.

There are a number of methods for probabilistic cost analysis. All have advantages and disadvantages; none are perfect. Some of the more widely used methods are described below. In these descriptions, we refer to the estimate excluding contingency as the "base estimate," and when contingency is added it becomes the "budget estimate."

The Monte Carlo Simulation

In the Monte Carlo Simulation (see Figure 10.10), the user defines equations which duplicate or model the way the estimate has been done. For example, the cost of excavation may have been estimated using this equation:

Cost of excavation labor = (cubic yards)(man-hours/cy)
(cost/man-hour)

The terms on the right side of the equation might be called "cost variables." A probability distribution is then defined for each cost variable, by specifying the maximum and minimum values that the variable could experience. A frequency distribution for the value of that variable can be deduced. This process is repeated until the entire

— User defines equations which "model" the estimate

   e.g. + cost of labor = (cubic yards) (man hours/cy)

      (cost/man hours)

— For each variable:

   Min   Estimate   Max

— Random-number generator selects value for each

   variable, calculates total project cost

— Process repeated "n" times to generate cost vs.

   probability curve

FIGURE 10.10 The Monte Carlo simulation. In this simulation, the user defines equations which model the estimate. A random-number generator than selects values for each variable, and calculates the total project cost. This process is repeated perhaps 1000 times to generate the cost vs. probability curves.

cost estimate is represented by the cost model, and each cost variable
has been assigned a maximum and minimum value.

The next step in the method is to simulate many possible outcomes
of the project, using the cost model and the frequency distributions for
each variable. This is done by selecting a random number (as one
might spin a roulette wheel in Monte Carlo) which is used to select a
value for each cost variable, according to its frequency distribution.
When a value has thereby been selected for each variable, a possible
outcome of project cost is calculated. This is one datapoint on the
frequency distribution curve for total project cost.

The process of selecting a random number, selecting a value for each
cost variable, and calculating a possible outcome of project cost is
repeated hundreds or even thousands of times, until a complete probability distribution of total project cost is generated. This distribution
can then be used to calculate accuracy, risk of overrun, and contingency requirements. The Monte Carlo simulation is generally computerized, and there are a number of good programs available which are
relatively easy to learn and use.

Hand Calculations Using Statistical Approximations

In those situations in which it is inconvenient or inappropriate to use
a computer program, there are a number of statistical-analysis techniques which can be done manually. These can also give an
approximate result which will prove useful.

PERT approximations

One of the best-known statistical approximations is provided by PERT
(Program Evaluation and Review Technique; see Figure 10.11). Used
primarily for scheduling, PERT gives us the following formula:

$$\text{Expected value} = \frac{\text{minimum} + 4(\text{most likely value}) + \text{maximum}}{6}$$

The expected value can be calculated for each cost variable or component of the estimate, and then the expected value of total project cost
is simply the sum of the expected values for each cost component.
Contingency is then the difference between the expected value of
project cost and the base estimate.

Cost-loaded network models

Some projects are planned and controlled by the use of cost-loaded
networks, in which the cost of each activity is identified, consistent
with its duration and resource requirements (see Figure 10.12).
These cost-loaded networks can be thought of as a form of project
model. One way that schedule and cost contingency can be addressed

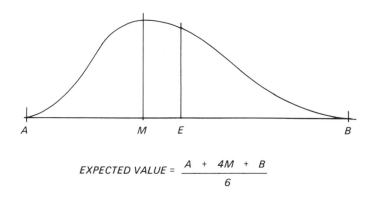

$$EXPECTED\ VALUE = \frac{A\ +\ 4M\ +\ B}{6}$$

FIGURE 10.11 PERT approximations.

in an integrated fashion is to define a model for which the duration and cost for each activity is at the 50% probability level. Schedule and cost contingency can then be defined as the difference between the end date and final cost predicted by the 50/50 model and the original, target model.

Approximating the standard deviation

The standard deviation is important, as it is the parameter which tells us the shape of the cost vs. probability curve and allows us to determine the probability of any cost level being overrun (see Figure 10.13). We can approximate the standard deviation with that formula. If we use this formula to calculate the standard deviation for each component

FABRICATE
PIPING

— Cost assigned to each activity
— Network and estimate for cost-time "model"
— Cost and duration for each activity can be set at
    "most likely" or at "50/50" level

FIGURE 10.12 Cost-loaded networks, in which a cost is assigned to each activity, a network and estimate for cost-time model is developed, and cost and duration for each activity can be set at "most likely" or at "50/50" level.

of the estimate, and if the total project cost is the sum of the component costs, we can calculate the standard deviation of the distribution
of all possible project costs by using the following formula:

$$\text{Variance} = (\text{Standard Deviation})^2$$

$$\text{Variance (total cost)} = \sum \text{Variance (components)}$$

Decision-tree analysis

A "decision tree" displays the alternative courses of action or outcomes
which are possible at a certain point in the project. If a probability
and a cost impact is attached to each outcome, the most probable result
can be identified. Decision trees are a good tool for analyzing specific
risks, though they do not allow us to analyze the combined effect of
many different variations.

Expected-value analysis

This type of analysis is similar to the decision tree in that specific
risks and cost outcomes are identified. A probability is attached to
each one, and the "expected value" is calculated as the product of the
probability and the cost outcome. In some applications, these expected
values are summed, to get the expected value of the total project.

Utility theory

All of us have a certain bias toward risk. We may have a high willingness to accept risk (the gambler mentality), or an aversion to risk.
Since all capital projects involve some consideration of risk and return,
utility theory may be used to calculate the risk/reward profile of the
corporation and key managers. This can then be used as a guideline
in evaluating projects. Utility theory provides a supplement to the
other risk analysis methods.

Standard deviation (σ) describes "spread" of probability
curve

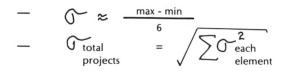

$$- \quad \sigma \approx \frac{\text{max - min}}{6}$$

$$- \quad \sigma_{\substack{\text{total} \\ \text{projects}}} = \sqrt{\sum \sigma^2_{\substack{\text{each} \\ \text{element}}}}$$

FIGURE 10.13 Useful approximations of standard deviation. The
standard deviation (σ) describes the "spread" of probability curves.

Pitfalls to Avoid

One common error in dealing with probabilities involves the basic concept of combined events. The probability that events A *and* B will
occur is the product of the probabilities of each event occurring, that
is:

$$P(A \text{ and } B) = P(A) \times P(B)$$

If we wish to describe the probability that event A *or* B will occur, that
probability is the sum of the probabilities of each event, that is:

$$P(A \text{ or } B) = P(A) + P(B)$$

Since some project risks are considered to act independently, and
others usually act in concert with other risks, it is important to
consider when probabilities or expected values should be added or
multiplied.

The Problem of Independence

All of the techniques described above require that the individual cost
variables or cost components be independent of each other, that is,
that variation in one variable be totally unrelated to variations in
others. While these methods give satisfactory results, it is, in fact,
unlikely that the requirement for independence between variables will
be met. It is therefore important, when designing cost models for
use in probabilistic analysis, to try to define cost variables or components such that they are as independent as possible.

It is also possible to develop equations for calculating the most likely,
median, and standard deviation for total project cost, in which the
"covariance" between variables and components is reflected. The
covariance can be thought of as a correlation coefficient between the
terms.

ADMINISTRATION OF CONTINGENCY

Once a project is approved and work begins, cost control becomes the
prime consideration. Contingency, being a major component of the
budget and control estimate, must be administered and controlled like
any other cost account.

Effective cost control requires that a cost forecast be made which
reflects the outcome which is foreseen at that time. The contingency
which is forecast to be required to complete the project should reflect
the uncertainties existing at that time, just as we saw in preparation
of the budget estimate. Therefore, reductions in contingency can be

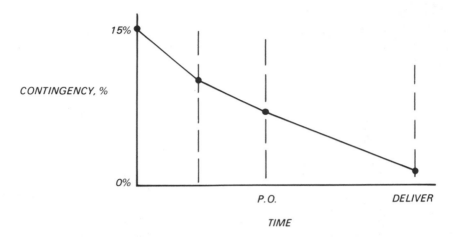

FIGURE 10.14 Administration of contingency (contingency = forecast
of remaining uncertainty).

planned. Major reductions in contingency requirements can be made
when milestones are reached, with the resulting major reductions in
uncertainty.

Project contingency can be allocated to major cost codes, according
to the uncertainty associated with each one. For each cost code, a
schedule is prepared, showing the major milestones in uncertainty
reduction as illustrated in Figure 10.14. For example, placement of
major purchase orders results in considerably greater confidence in
the forecast of escalation. At each milestone, we can define the conti-
gency that will be required to cover the remaining uncertainty. Between these milestones, we draw a straight line. We now have a con-
tigency rundown plan which can be used as a guide when setting
contingency for cost forecasts, as well as a basis for monitoring
contingency funds.

A key point in contingency administration is that *the contingency
required to complete the project should always be determined by the
risks and uncertainties that lie ahead*, not by consideration of how
much contingency has been "used" to date and how much therefore is
"left."

CHAPTER SUMMARY

Uncertainty and risks are as much a part of project work as time and
material. Contingency is the way we deal with uncertainty and risk,
and it should be addressed in the same way we deal with any other

project-management problem: firmly, clearly, and with the best technology available. Although the available methods are not perfect, they can go a long way toward improving current practices.

Because there are, in general, negative emotions associated with uncertainty and risk, project engineers should take the lead in establishing the terminology and methods required to assure that investment and project-management decisions are made with proper recognition of the financial risks involved.

Part V
Computer Applications

11
Computer-Assisted Project Management

WHY CONSIDER COMPUTERIZATION?

Computers have become such an accepted part of our everyday lives that it becomes easy to lose perspective on the fundamental issues of why they are appropriate in a given situation, what they should and shouldn't be expected to do, and how to achieve the computer capability we need. These questions have become markedly more complex in recent months as the proliferation of mini- and microcomputers has offered many new possibilities while raising new questions.

Fortunately, in spite of the many new developments in computing, the fundamentals of system design, development, and implementation have not really changed. Equipped with these fundamentals, we can use the latest computer technology in an effective way, regardless of how rapidly or dramatically the technology changes.

This chapter covers the fundamental concepts of system design, development, and implementation as they apply to project management. Like most other chapters in this book, this one is not intended to make the reader an expert. Rather, it is intended to give a project engineer or manager all he needs to know to be able to assure that effective computer-assisted project-management techniques are available for his projects. We will, for that reason, deliberately retain the perspective of the computer system *user* throughout.

Our first consideration as users should therefore be: "why consider computerization in the first place?" If we look at the most effective computer systems, and consider inherent capabilities offered by computers, we find some or all of the following general conditions for effective information systems:

A considerable amount of data which needs to be handled
A requirement to file data and also present selected data when needed
A short time available to process data
A relative simplicity and repetitiveness of operations for handling
    data
A large number of people in the organization involved with the data
    going into the system, the information presented by the system,
    or both
The consideration that the availability of timely and accurate
    information is essential to the achievement of business goals.

It is certainly evident from the above that the small-project management situation fits all of the criteria for effective computerization. We
have a lot of data to handle in a short period of time, a lot of people
involved, a need to be able to store data for later use, a lot of data
needing manipulation using simple and repetitive calculations, and,
most important, a need for timely and accurate information. As we have
seen in the preceeding chapters, computer power can be a time
saver in:

Network analysis and scheduling
Resource analysis
Cost estimating and expenditure forecasting
Management reporting
Decision making

So computer-assisted techniques deserve close consideration in the
small-project environment.

Due to the fast-changing world of computing, many people look at
the option of implementing a computer system and say "I think I'll wait
until the prices come down even more before buying. If I buy something now, it will probably be obsolete even before it's fully paid for!"
This is a understandable reaction, particularly since prices will probably continue to drop while the quality and quantity of computing power
sold will continue to rise. Of course, the only response to that is to
point out that the world is changing rapidly due to the broad acceptance and use of computing and we can hardly afford to be left behind,
as individuals or as organizations. And, this rapid acceptance is positive proof that the improvements brought about by computerization are
real. With mini- and microcomputers as inexpensive as they are, the
price factor is not as critical as it once was, and today's computers,
even when obsolete, will continue to provide useful functions. So,
let's plunge bravely into the world of computing, beginning with the
definition of terms.

DEFINITION OF TERMS

As in other subjects covered by this book, it is worthwhile to define
the terms we will be using. As before, these definitions are the ones
that the author has found to be commonly accepted.

Project-management system: A computer-system for processing project data in order to facilitate project-management actions.

Project-management method: A series of specific actions to accomplish
a project-management task. The method may include use of a project-
management system. The method is documented in a procedure which
is followed uniformly.

Data: Numbers and words which represent some aspect of the project. "Raw" data is in the form in which it was captured, "processed"
data has been subjected to some manipulation.

Information: The presentation of processed data and written materials in a way that conveys specific knowledge which will enable the
person receiving it to perform project-management functions. Information can be presented in printed reports, with graphics, or on a
screen display.

Input: The entry of raw data into the system. Input may be by
means of a keyboard, magnetic tape or disc, punched cards (now
somewhat out-of-date), or electronically through a communication
network.

Output: The presentation of information prepared by the system.
Output can be in the form of printed reports, management graphics,
a screen display, an electronic message to another computer, and/or
a file on a disc or tape.

Random-Access Memory (RAM): The part of a computer in which
data is processed. The amount of RAM indicates the size of the program
and the amount of data that can be handled in memory.

Read-Only Memory (ROM): That part of the memory which contains
the instructions to the computer. It cannot normally be accessed or
changed by the user.

Bit: A measure of the quantity of data stored or processed. One
bit is one yes/no signal.

Byte: A measure of the quantity of data stored or processed. One
byte equals a word consisting of eight bits, and is sufficient to describe a character or number. A *word* is the number of characters
that the computer can process together. Many current machines have
"eight-bit processors" meaning that they process information in terms
of words of eight bits. More powerful machines for project applications
have 16- or 32-bit processors, and are therefore able to process more
information in less time.

Hardware: The combination of copper, steel, plastic, and silicon
that makes up the physical part of a computer. Hardware alone cannot
perform any functions.

Operating system: The programming which is built into a computer that enables it to function. The operating system "tells the hardware what to do," and how to process data with it's memory, discs, keyboard, and other devices. The names of most operating systems contain the words "Disc Operating System" (DOS). It also tells the computer how to interface with the various items of software which will be run on it. The operating system is important because it defines the type of software that can be used, as well as the type of programming that can be done.

Program: A set of instructions for the computer to follow. A program defines the input, processing, and output that must be performed. Programs are written in various languages such as BASIC, C Fortran, and APL in which the instructions are expressed in a way which can be interpreted by the operating system.

Software: A package of one or more programs which is designed for a specific application. Software will work with one or more operating systems and hardware configurations. Software can be sold commercially as a stand-alone product for general use. Specific-application software may be purchased, written as a new program, or, in some cases, as a combination of both.

Bundled system: A computer system in which hardware and software are sold together, in one integrated package. In a bundled system, the hardware and software cannot be separated.

Unbundled system: A computer system in which hardware and software can be purchased separately. The software is said to be "machine portable" in that it will usually run on more than one type of hardware. One advantage of an unbundled system is that it enables the purchaser of the hardware-software package to use the hardware for other applications, or to purchase only the software.

Timesharing: A means of accessing a computer which is owned by someone else and using only as much time as is needed. Timesharing can be obtained through a number of large companies offering such a service (e.g., McAuto, Boeing Computer Services, ADP). Timesharing can be accomplished by inputting data at the local office of the timesharing company, or by using an input/output terminal in the office.

Network: A means of linking together several computing components, such as a large mainframe computer, several terminals, and different software packages. Networking can be accomplished with special hard-wiring, use of telephone lines, satellite communication, or special networking packages including both software and hardware.

Terminals: Local workstations at which the system user can input data using a keyboard and receive output in the form of a visual display and/or in print. A "smart" terminal (usually a microcomputer) is also capable of storing and processing data itself. It can, therefore, "upload" data which it has stored and/or processed to another computer

or to other terminals, as well as "download" data which it receives
from another computer. A "dumb" terminal is simply an input/output
device and has no internal storage or processing capability.

Mainframe: A large, powerful, centralized computing facility supporting most if not all of an organization's data-processing needs.
Characterized by high-speed processing (e.g., millions of instructions
per second), extensive storage capability (e.g., billions of characters),
multiple input and output facilities, a wide range of software, complexity of operation far beyond the skills of the average user, rigid environmental requirements (e.g., temperature, humidity), and very
high cost. The organization with a mainframe computer will inevitably
have a major organization function (Data Processing, Management
Information Systems, etc.) which is responsible for servicing the
various users.

Minicomputer: A powerful computing facility which can support all,
some, or perhaps only one of the organization's data-processing needs.
The minicomputer is still a sophisticated system, but it is apt to be
much more so in the hands of the user rather than in a data-processing
group. Most "minis" can function in any reasonable environment, are
portable, and far less expensive than a mainframe (typically $50,000-
$250,000). Like mainframes, current minis can support distributed
processing (examples: Hewlett Packard HP3000, Digital Equipment
PDP VAX, and Prime).

Microcomputer: A small personal computer dedicated to one or more
relatively simple computing tasks. Because the micro is almost entirely
in the hands of its user, and requires little expertise or training, if offers the benefits of convenience and responsiveness which larger, more
complex systems cannot match. The very low cost of the micro (e.g.,
$5000) not only makes it easy to implement, but also means that the
risks of developing a system which becomes obsolete or simply doesn't
work well are small, as the worst that can happen is that we scrap a
small computer and try again. It is because of these features of low
cost, user convenience, and ease of implementation that micros are a
good candidate for small-project applications (examples: Apples II and
III, IBM PC, Radio Shack TRS 80, etc.).

Home computer: Even smaller and less expensive than micros, home
computers are intended for education, games, and simple home applications. They are included in this list to indicate that, when we discuss
micros, we are not including home computers.

Peripheral equipment: Devices which are external to the processing
portion of the computer and provide input of data, output of information, storage, and special functions such as graphics.

Interactive system: A computer system in which the user interacts
in real time with the computer. This is usually done with a terminal
in which commands or questions are processed quickly and a result
displayed. In comparison, the "batch" type of processing takes all the

input at once (usually in the form of punched cards) and, later, delivers the printed output.

User-friendly system: A system which enables the user to interact easily with the computer, with simple commands. User-friendly systems typically feature Englishlike commands instead of the usual coded commands; "on-line tutorials," in which the system instructs the user as he goes; and helpful diagnostic messages when the user makes an input error. User-friendliness is extremely important in the small-project environment, since most users are not going to be skilled computer operators.

Menu-driven system: An interactive system in which the user is presented with a series of options from which to select, and then follows the instructions (or "prompts") as displayed on the screen. In effect, the computer is telling the user what to do. Menu-driven systems are relatively easy to learn and use and are appropriate for users with little system training. The disadvantage of the menu-driven system is that it is relatively inflexible, and this can frustrate the skilled user and prevent full utilization of the system's capabilities.

Command-driven system: An interactive system in which the user has a variety of commands which he can employ to tell the computer what to do. This type of system is more flexible than a menu-driven type, and enables the user to obtain exactly the results he needs. However, it also requires more sophisticated computer knowledge on the user's part. Many systems are available which have both menu-driven and command-driven capability, and this is very useful in the small-project environment. For example, data from timesheets could be inputed by a clerical person with little training using a menu-driven routine, while decision analysis and cost and schedule forecasting were done by a project engineer using a command-driven routine.

CONSIDERATIONS IN THE DESIGN OF PROJECT-MANAGEMENT SYSTEMS

In this discussion of project-management systems, it should be kept in mind that a computer cannot manage a project. What it can do is make the people responsible for the project as effective as possible, and, if we reflect on the special problems of small projects as discussed in Chapter 1, we can see that there is, generally, a lot of room for improvement.

As we saw in Section III, the essence of project control boils down to the timely processing and presentation of useful information and was summarized as follows.

A project-management system is a means of presenting:

The right information
In the right format

To the right person
At the right time

The project-management functions which benefit from computer assistance include:

In the project-evaluation phase:
    Cost estimating and expenditure forecasting
    Planning and scheduling
    Resource planning and analysis
    Economic analysis
    Analysis of risk and contingency
    Bid preparation and analysis
    Management decision making
    Collection of project data
    Design and technical analysis
    Document control
    Contractor and bid evaluation
In the project-execution phase:
    Cost control
    Progress measurement and schedule control
    Productivity analysis and control
    Materials management
    Control of labor and equipment resources
    Cost and schedule forecasting
    Contract administration
    Document control
    Quality control
    Risk management
    Field supervision
    Management reporting

To build a project-management system for a specific application, all we really have to do is combine the basic elements of system design to suit our requirements. These elements are:

Input in the form of keyboard entry, data from a storage disc or tape, or communication from another computer.
Processing or data manipulation using equations and a programmed procedure.
Storage of data or results on a tape or disc.
Output in the form of displays, reports, and/or graphics.

As project engineers all we really need to know about system design is how to define and combine these four elements of a system, such that the computer does exactly what we want it to do. Our system design becomes a specification which will be satisfied either by

purchased software or by a program which we will have written. The
software will then be mounted on some hardware which always fits our
specified requirements.

However, before a system can be designed, it is necessary to address some very fundamental questions about the work we do involving
how we do it now, how we should do it, and how we could do it with
computer assistance. In other words, the essence of system design
is not programming, nor is it selecting the hardware and software. It
is defining the job that the system is to do. To do that, we have to
understand our own job very well.

BASIC STEPS IN SYSTEM DESIGN
AND IMPLEMENTATION

To those of us involved in project management, it may seem like a lot
of unnecessary work to follow all the system design steps shown below.
After all, all of the system vendors claim that their product will solve
all our project-management problems, so why not just go ahead and
pick one? The number and diversity of available computer systems is,
in fact, one of the reasons why the preparation of a system's specification is so important. The only way to cope with the volume of information, the pressure of time, and the intensity of salesmen is to
have a clear idea of exactly what is wanted.

The system's design is also important because projects exist within
an organization, and the power of the organization is mightier than any
computer. Therefore, if our system is to be effective, and it it is to
make our hard-won project-management methods effective, it must fit
the organization and enhance its operation. To assure that this is
the case, we must know ourselves, our jobs, and our needs—as well
as those of the organization in which we work—before we can know
what computer system will work for us.

The basic steps in system design are as follows:

1. Define user requirements
2. Develop system performance specification
3. Prepare plan and cost estimate for implementation
4. Obtain management approval to proceed
5. Prepare sample project and reports
6. Contact suppliers and develop a short list
7. Conduct "compute-off" and evaluate results
8. Obtain management approval for implementation
9. Implement system

Each of these steps is described in detail below.

Step 1: Defining User Requirements

This first step is both the most important and the most difficult. It requires a thorough analysis of current and future job practices as they relate to the receipt and manipulation of data, and the presentation of information. The "primary user" referred to below is the typical person in the organization function for whom the system is designed (e.g., the project engineer). There are, of course, other people in other functions who might also be users; we will refer to them as "secondary users." Taking the perspective of the primary user, we can develop the necessary definitions by considering the following questions.

Review and Definition of Existing Practice

Who will be the primary user of the system?

What are the objectives of the primary user's job function? What methods are now used to perform that function?

What information or data does the primary user get now? From whom does he get it, and in what form?

What information does the primary user provide to others? What information is provided to whom? To what use is it put?

What data manipulation is now performed by the primary user in order to provide the necessary information to others?

What problems exist with current practice?

What improvements could be made to the current practice?

Who would the secondary users be, to enhance current practice? What are their information needs? What information must they provide to others?

Definition of Ideal Practice

Can or should the primary user's job function be changed? If so, what should it be?

What methods should ideally be used?

What information does the primary user need to do his job ideally? From where should he get it, and in what form?

To what organizational functions should the primary user provide information? In what form should it be provided? To what use would it be put?

What data manipulation should be done by the primary user to provide this information?

Who would be the secondary users in the ideal situation? What information would they receive? What would they provide to others?

What improvements would the ideal practice achieve over current practice?

Specifying Short-Range System Requirements

What existing user functions could be computerized for greater
   efficiency?
What immediate improvements could be gained by computerization?
What existing computer systems might be used?
With what existing computer systems might the new system interface?
   What would be the type of data exchanged across the interface?
What level of computer capability will the primary user have? What
   other levels of user capability need to be accomodated?
What organizational functions will be putting in what data?
How much data needs to be stored? How much of that needs to be
   readily accessible for calculations and analysis?
How fast do the calculations have to be done? Can we, for example,
   use a smaller computer which takes a full day to process and print
   reports, or do we need a larger one which can do it in five
   minutes?
What systems exist for coding project information? If new coding
   systems will be developed, what constraints (e.g., number of
   digits) may be imposed by existing practices?
How much flexibility is required in the system's operations? What
   operations need to be flexible and to what extent?
What cost limitations will apply to the design and implementation of a
   computer system? What schedule limitations apply?
Of all the specified features, what is the priority of each one? For
   example, is price more important than storage capability? Do we
   prefer buying a software package "off the shelf," or designing
   our own? Can we separate our "wants" from our "needs" in terms
   of system requirements?

Specifying Long-Range System Requirements

Specifying requirements requires a reevaluation of the issues outlined
in our discussion of short-range requirements, but with a view toward
the future. Any system we develop must be able to meet future as well
as current requirements, or be adaptable to newer systems as they
become available. So, for example, if the workload of the project-engineering department is expected to double in the next two years, we
must be sure that our system is expandable to that capacity. The company may change or add to its computer systems, so any systems we
introduce will have to be compatible with the new system. New codes
of accounts, procedures, and interfaces may also be implemented.
   One of the key aspects of system design is the identification and
evaluation of "wants" versus "needs." Just as in any design function,
there is a tendency to include capabilities that are desirable but not
really necessary. The problems of distinguishing between wants and
needs is particularly difficult in system design, since so many of the

benefits have to do with "managing better" and are therefore difficult to quantify. One way to keep the system's specification under control is to specify the minimum system that will do the minimum job, and then put each added feature or capability through a cost/benefit analysis, just as one would do for a proposed design change.

Step 2: Developing System Performance Specification

The previous step was aimed at determining what we want the system to do for us. This step involves describing the specific characteristics of the system which will have the capabilities we need, both now and in the future. The system design consists simply of describing the four basic system elements: input, processing, storage, and output.

System Input Specification

In this step, we specify the type of input, the amount of data to be inputed, the sophistication of the person making the input, the location of the input function, the flexibility required, and the need for interaction with the main system during input.

The input to the system can take the form of interactive terminals, punched cards, or electronic data transfer systems such as discs or tapes from other computers, or through electronic communication.

If input is to be in the form of interactive terminals which, in most cases, fits the small project environment, we have a number of options:

"Smart" terminals which are actually small microcomputers and provide local computing capability. Such terminals can process input data so that it goes into the main system in a certain format and the user can perform certain functions at the terminal without having to access the rest of the system. When the terminal performs some processing and then inputs the result to the main system, it is said to be "uploading" information into the system. When the main system performs some processing and then sends the resulting information to a terminal, it is said to be "downloading."

"Dumb" terminals, which are strictly input/output devices for the main system.

"Local area networks," in which a number of computers and software packages are linked together and able to interact with each other. Most of the major manufacturers offer this capability.

Menu-driven input and/or command-driven input.

Remote terminals with tie-in to the main system via telephone, telegraph, or other forms of telecommunication.

An example of input by data transfer is the use of timesheet data which is already input to the accounting system. We might arrange for the man-hours in each labor category to be transferred to a tape or disc which would then be an input to our cost control system.

System Processing Specification

In this step we must specify what the system must do with the input data: that is, what calculations are to be performed, how much data needs to be handled, how many calculations need to be made, how quickly the calculation must take place, what data files need to be read, what comparisons are necessary, and what results are desired. It must also be recognized that an interactive system allows the user to follow and even to affect the data processing function and we there-fore need to identify those calculations which are always the same and require no interaction from those for which interaction is desirable. For example, the preparation of weekly expenditure reports would, in general, not be an interactive process, whereas the routines used for decision making would be. Therefore we need to specify the operations which are to be interactive, and the extent of interaction desired.

The system processing functions which are most often used in project management are: arithmetic calculation data retrieval, comparison, and extrapolation.

Calculation functions are performed the same way we would prepare a progress report, an estimate, or any other simple task. As we have seen in previous chapters, much of the work involved in progress-report preparation is the calculation of current figures and the compar-ison of those figures with the budget plan. To do this by computer, we retrieve the file of budget data and perform comparison of actual data vs. the budget plan. We have also seen in previous chapters that extrapolation of current trends is a useful way to highlight potential problems and this is another function which is easily computerized.

These functions are generally provided by the purchased software. One feature to look for in software is the flexibility to define the func-tions that are to be performed. For example, there are many schedul-ing packages which do network analysis, but they all vary in terms of the user's flexibility in defining the coding system, to select arrow or precedence notation, to define hammocks, to load resource and cost data onto the activities, etc.

Another important feature to look for in software is the ability of one software package to interact with another. For example, in proj-ect management we are interested in network and resource analysis, cost estimating and forecasting, management reporting, word proces-sing, management graphics, and database capability. Although most systems have software to do each of these functions, we may find that the ability to tie them all together will vary between systems. If these functions can be integrated a much more efficient system will result.

System Storage Specification

In this step we must specify the amount of data to be stored, the format for storage, and the requirements for recall. For example, we might wish to construct a database of norms for estimating, derived from past projects. This database would be carefully designed and probably integrated with an estimating routine. It would probably be stored on a disc or tape in order to be readily available. In contrast, we might also wish to store data from past projects for future analysis. This data could be in a form of "dead" storage.

System Output Specification

The success of system implementation depends, to a great extent, on the effectiveness of the output. If a system produces clear, concise, useful, and easy-to-read reports, it is bound to be a success. Conversely, many good systems fail to win the acceptance they deserve, because the users simply cannot relate well to the output produced. This principle is particularly useful in the small-project environment where, as mentioned earlier, we are trying to present "the right information, in the right format, to the right person, at the right time."

There are three types of output which are relevant to small projects: printed output, visual displays, and graphics.

Printed output

The reports generated by printed output will probably be circulated throughout the company to advise managers of status and problem areas. They should permit management by exception and provide only that information which is necessary to the particular report recipient. In order to achieve this, a system feature which is most useful is the flexible report writer. This feature enables the design of many reports, each of which has a separate format and which displays specific data. Systems without this feature have one or more report formats which cannot be changed.

Another useful feature in report preparation is the ability to sort, select, and order data. This makes it possible for us to select the specific information we want for each report, sort it into appropriate categories, and print it in a given order. For example, a manager's report listing all projects might be designed such that the project with the largest cost overrun is printed first, and subsequent projects ordered by decreasing overrun amounts. From our array of project data we might select cost data for an accounting report, and order it by cost code: the same data might be arranged by contract number for a report to the contracts engineer.

Visual displays

The image that an interactive system makes on the screen is, of course, only important to the user. However, visual displays are an important form of output because the user's effectiveness depends to a great extent on the system's ability to give quick outputs of useful information. One useful feature of some new systems is the ability to present several displays on one screen. Another feature to look for is the display of messages which tell the user what is going on when the computer is calculating, and also what is going wrong. Finally, we would like to have a visual display that makes it easy to see the results of a change to the project parameters. For example, we might be varying logic and durations to test the effect on the end date. We would, therefore, like a display format which makes it easy for us to see the end date after each run.

Graphics

Nowhere is the principle that "a picture is worth a thousand words" better illustrated than in management graphics. With well-designed graphics, the meaning of an otherwise confusing and uninteresting set of figures can be driven home clearly, quickly, and forcefully. And, as we saw in Section III, a great deal of the effectiveness of our project-management effort will hang on our ability to focus the attention of busy managers to those areas in which it is required. So management graphics deserve consideration in the small-project environment. Most computer systems for project management either have such a capability or are compatible with graphics software.

To specify the graphics capability required, we need to identify the type of graphics, the quality required, the source of data to be used, and the quantity required. For project management, the most common graphics types are:

  Barcharts
  Networks
  Manpower histograms
  Organization charts
  Piecharts
  Tracking and trending curves

The quality determines the type of drawing equipment required. For example, if top quality four color graphics are desired, then a four-pen plotter with associated software is needed. This, however, can add significantly to the system cost. An alternative might be to forgo certain graphics functions, and prepare barcharts, histograms, and even tracking curves on the line printer.

Step 3: Preparing the Plan and Cost Estimate
for Implementation

At this point, we have an idea of what we are trying to accomplish with computerization, and what kind of system will be required. Before going further, it is appropriate to exercise some project-management techniques on ourselves, and establish a plan and estimate for what is to come.

The major cost items that can be expected are as follows:

Purchase or lease of hardware (i.e., computer and peripheral
  equipment)
Purchase or lease of software
Purchase of accessories
Maintenance contract for hardware and/or software (if applicable)
Modifications to office facilities to accomodate computer (if applicable)
Time required for training and implementation
Supplies (computer paper, discs, etc.)

The computer acquisition and implementation should, in fact, be treated as a project just like any other, with a budget and a schedule. It should be recognized that the estimate prepared at this stage will be preliminary, in that we have not yet gone into the marketplace and obtained quotations. However, it is important to prepare a preliminary budget at this stage in order to accomplish the important next step: obtaining management approval to proceed.

Step 4: Obtaining Management Approval
to Proceed

The purpose of this step is to advise management of what we are doing, obtain their general approval of the overall plan, and specific approval to proceed with system evaluation and recommendation. This step is suggested for three important reasons:

1. We wish to have company management "on our side" in this endeavor, since the organizational implications of a new computer system tend to be far greater than the capital cost might indicate. We want to be sure that management understands and supports the objectives of better project management, towards which the computer is merely a means.

2. We wish to impress upon company management that the computer system is intended to be a cost-saving device, which is to be evaluated and managed like any other small project. Far too often, engineers and managers decide first that a computer is needed, then decide what kind of computer to buy, and never subject these subjective decisions to the same sort of scrutiny that any other investment decision would endure. We must never lose sight of the fact that there is only one good reason

for computerization: that it will be profitable. We want our management to know that we are operating that way.

3. By providing preliminary cost/benefit figures, we are in a position to review our plans with management, and perhaps identify some changes. For example, we might be pleasantly surprised and find that management is enthusiastic about the project, and encourages us to specify and implement a more ambitious system than we would have done on our own. And, since small-project management affects so many parts of the organization, there may be implications and interfaces that only management are aware of, and that insight is an important part of a successful system.

Once we have obtained approval to proceed further, we are ready to journey into the bewildering world of modern computing. But before we are swept up in electronic excitement, we need to prepare ourselves for this journey. As always, the best way to be prepared is to know exactly what we want.

Step 5: Preparing the Sample Project and Reports

Although we went to a great deal of trouble in Steps 1 and 2 to define our needs, we had better expect that there will be a lot of systems that will meet or exceed them. We had also better expect that we are likely to see a bewildering array of computers, salesmen, and impressive demonstrations, at of the end of which we are likely to say, "But they all seem so good. How can I possibly choose one?"

The truth is that all the systems are probably very good. Each has features which its proponents claim make it immeasurably better than the others. The only difference between them will be that some will suit our needs better than others, and one will suit us best of all. But how will we know when the right one comes along?

The answer is to prepare a sample project (or projects) which represents the typical situation which our system will have to handle. We should also make up some input data in exactly the form in which we would normally receive it and design some management reports that show exactly what we would like the system to produce. Now, instead of approaching our friendly computer salesman and saying, "Show me what your system can do," we can say, "This is what I need, show me that your system can do it!"

We have therefore defined system requirements, system performance specification, and a sample project for use in testing and evaluation. We know what we want and have expressed it in specific terms. At the same time, we are free to change any aspect of our system design as we become aware of new system features and capabilities. We have also established a basis for fair testing and comparison of the systems selected for in-depth evaluation.

Step 6: Contacting Suppliers and Developing Short-List

The purpose of this step is to identify three of four systems which are worthy of in-depth evaluation. This process is currently made more difficult by the tremendous number of hardware and software packages relating to project management (easily over 100 at the time of this writing). There are, however, a number of sources of information about available systems:

1. The Project Management Institute conducts a biannual survey of project-management software, which is available from their office in Drexel Hill, Pennsylvania.
2. The American Association of Cost Engineers also accumulates information on project-management systems which can be obtained by contacting their office in Morgantown, West Virginia.
3. Reference catalogs are now available listing diverse products and services.
4. Trade shows and technical journals carry a profusion of advertisements for project management systems.
5. Many special seminars are now available throughout the country on computer applications.
6. Most major cities have a profusion of companies offering computer sales and services. These include stores selling microcomputer systems (an excellent place for a project engineer to spend a lunch hour or two), small companies selling software, and large companies selling a variety of solutions including timesharing, and micro- and minicomputer systems. One can even learn a great deal just by using the Yellow Pages and asking around.

When conducting this "survey" of what is available, it is well to bear in mind these few hints:

1. *Consider software before hardware.* The effectiveness of a system will be determined first and foremost by the software: if the programs meet our needs, then it is likely to be relatively simple to arrange the hardware as necessary. Although, as engineers, we like to look at the hardware spec sheets and compare the esoteric data we find there, that type of evaluation is appropriate only in the final stages. To put it bluntly, software is everything in a computer system.
2. *Take the time to read the user's manual.* Short of running the software, there is no better way to get a feel for what the software can do than to read the user's manual. The clarity and completeness of the manual is itself a good indication of the quality of the software package.

3.   *Avoid writing new software.* Many people jump too quickly to the conclusion that their needs can only be met by writing their own software. This may be true in some cases, but, in most project-management applications, there are few functions which cannot be satisfied by some existing software package. Although the existing package may represent a compromise, it must be remembered that the time, cost, and risk are greatly reduced. Software-development projects are notorious for overrunning budget and schedule by several orders of magnitude.
4.   *It pays to be skeptical.* For example, one should pay no attention to claims that a certain capability we require will be available "on the next release of the software." We are interested (at least initially) only in that which is available and demonstrable now.
5.   *Insist on talking with someone who is knowledgeable in the specific area of interest.* With the sudden proliferation of computer stores and companies, combined with the proliferation of hardware and software products, we could hardly expect an equal proliferation of experts. It is, of course, impossible even for a real expert to be familiar with every application of every system and, as a result, good advice is hard to find. So, when discussions progress to the point where it is possible to identify the items in the hardware or software product line which are of interest, it is worth the extra trouble to find and talk to the person who is most knowledgeable about those specific items.
6.   *Examine closely any claims about hardware and software interfaces.* Experience has shown that the problems of getting machines to communicate with each other and the problems of linking software are often far greater than anticipated.

At the end of this step, we should have identified three or four systems for in-depth evaluation. More candidate systems will only add to the time and cost of the evaluation process, and will probably not increase the likelihood of the correct system being chosen. We may, during this step, also have made some revisions to our system design, to reflect what we learned. Now we are ready for the "try before you buy" part of the project.

Step 7: Conducting the "Compute-Off" and Evaluating the Result

The purpose of this step might be compared to selecting a new car by driving each of the cars selected for possible purchase around the same course and evaluating the results according to a set rating sheet. In this case, our "course" is the sample project, data, and reports.

Requesting the Demonstration

We begin by asking each selected supplier of hardware or software to set up a hardware/software configuration identical to that which we are considering, input our sample project network and data, and then demonstrate how the system will accept data and provide output in our format. We give him plenty of time (say a week) but insist also that we be able to operate the machine ourselves during the demonstration. We also make him aware that we are evaluating other systems in the same way, and will be making a firm decision to purchase based on the results.

Of course, the salesman's willingness to do this may depend to some extent on the size of the potential sale, but buyers should not be reluctant to ask for such a demonstration. Most good salesmen of reputable computer products will be delighted to have a customer who knows what he wants and can recognize good results, and they will be pleased to provide the custom demonstration as requested. If not, that in itself is a good indication of the kind of support that can be expected after the sale.

Preparing the Rating Sheet

The categories on the rating sheet, and the relative weight given to each one, are determined by our definition of the system's design requirements. However, we can identify some system characteristics that will inevitably be important for a project application.

1. *User-friendly interaction*, i.e., the system should be easy to use, with a low "frustration factor." User-friendly features include English-type commands (as opposed to letters or numbers), tutorials (in which the system actually guides the user along), helpful error messages (e.g., "system will not accept decimal input" is more helpful than "syntax error") and help commands (in which, when in doubt, the user simply types "help" for a complete listing and explanation of the commands available at that point).

2. *Flexible reporting format*, or a fixed format which is well-suited to the particular application.

3. *Good interface characteristics* between the hardware and any existing hardware in the company. For example, it is a plus if our new microcomputer is capable of communicating with the existing mainframe computer even if we have no intention of doing so at present.

4. *Good interface characteristics between software packages.* For example, can we pass data and information from one software package to another?

5. *Good user support* from the hardware and software suppliers.
6. *Software documentation* providing a clear, easy-to-use user manual.
7. *The provision of training programs* for users.
8. *The ability to upgrade* hardware and/or software to increase capacity or to use newer products.

We should bear in mind that we are rating the software, the hardware, and the supplier in terms of price and performance.

Evaluation of the Results

To evaluate the results, we like to use a small committee of individuals with diverse skills, such as computing, project control, estimating, accounting, and management reporting. Each individual rates each system, and a composite score is agreed-upon by the committee. Evaluations should be made, at first, without regard to price. Once the performance ratings are established, then price can be considered and cost/benefit analyses performed.

Step 8: Obtaining Management Approval for Implementation

The purpose of this step is to present a specific proposal for implementation to company management, and to obtain approval for full implementation. As before, it is important to treat the subject of computerization objectively, and show in specific terms how profitability will be enhanced, due to time and cost savings on projects and in the project engineering and management functions. It is important that the first step in computerization be one which provides visible and significant benefits, so that the concept is well accepted.

Management support is even more important at this stage. During implementation, people are likely to react with cautious enthusiasm, tolerance, grudging acceptance, reluctance, and even outright hostility. We will be much better equipped to handle this if we, and everyone else, know that company management supports the program.

The proposal to management will probably be an update of the preliminary proposal, with greater details on the time, cost, and plan for short-range implementation, the possible longer-range developments, and a detailed economic justification. The organizational aspects of computerization will be uppermost in the minds of management, and that should also be squarely addressed by describing how the roles of existing functions may be changed, how lines of communication will be altered, and how efficiency in various departments will be improved.

Step 9: System Implementation

Setting up the First Project

After installation of our computer system, the next question is, "where to start?" Many companies find it useful to take a project and implement it on the new system while continuing to control it using the old methods. This provides a check on the results and avoids making a "guinea pig" out of the first projects. Often there are lessons learned from the early projects that indicate improvements that can be made before the system is implemented on a broad scale.

Keeping Everyone Informed

A key factor in the acceptance of the new system is that everyone must be kept informed, and we must display a willingness to share experiences with the new system and to help others obtain similar benefits. It is especially important to be sure that those individuals and organizational functions which interface with our new system be made to feel that they were consulted early enough to assure that their needs and constraints were considered.

Some of the methods for keeping people informed include progress memos, coordination meetings, and in-house seminars and training sessions.

Preparation of Procedures

It is important to assure that the new system is used in a manner which is consistent with the rest of the project's engineering and management functions. The best way to assure this is to prepare procedures which specify how data is input, processed, and output. The procedures should show how each organizational function interfaces with the system, as well as specify who is in charge of the system. Each function which will use the system should have specific procedures to follow.

Many systems have different levels of security, which are protected by passwords and other procedures, and the limits of each user therefore must be defined. For example, payroll or personnel data can be included in the system, but access to that data must be restricted. Similarly, we might construct an estimating database which is available to any project engineer for estimating work, but which can only be updated by those responsible for it. In that case, access to "read" from the database is unlimited, but access to "write" to the database would be restricted.

The procedure should also specify who the "computer czar" is to be, that is, the person who answers questions, resolves problems, sets priorities, listens to complaints, deals with the vendors, and generally coordinates system use. The recent surge in the use of personal computers in business has had a somewhat undesirable side effect of diluting the supervision and coordination of computer usage within a company. In past years, companies purchased a mainframe computer to service the entire organization, and users had to obtain the services they needed from the Data Processing (D.P.) Department. Because systems were expensive, maximum utilization was the goal, and users were often faced with a long wait for the needed capability to be provided by an already overworked system. In recent years, the microcomputer has enabled the user to implement his own solution in his own way, without recourse to the D.P. Department. While this has broadened the effectiveness of computers, it also has created many situations in which individual effectiveness has improved, but organizational efficiency has not. Therefore, today's responsible user has to assume some of the responsibility previously borne by the D.P. Department, to make sure that others are aware of what they are doing and that their system is compatible with others in the organization.

Training

Formal training courses are a significant aid in the acceptance and effective utilization of the system. The training course can be based around the user's manual, which describes how to operate the system, and the procedures, which describe how the system fits into company operations.

A REVIEW OF CURRENT PROJECT-MANAGEMENT SOFTWARE

Since most project-management needs can be satisfied by a variety of existing systems, it is worthwhile to review some of the characteristics of current systems. Project-management software can be divided into a number of categories as shown below. When implemented by a microcomputer, these programs are usually separate from each other, whereas larger systems can handle several of these functions in an integrated manner.

Network Analysis

A great many software packages are currently available to perform network analysis. Virtually all of these perform forward and backward pass calculations of critical path and float. Some of the features which vary between software packages are:

The size of the project handled (i.e., number of activities)
The ability to use precedence or arrow notation
The flexibility of the coding system used to identify activities
The ability to perform hammocking
The ability to calculate free float

The ability to perform resource analysis, resource aggregation, and
levelling

The number, type, and flexibility of resources permitted

The flexibility of displays and graphic output

The speed of analysis

The ability to handle multiple projects simultaneously

The ease and flexibility of updating (e.g., out-of-sequence updating
which allows updates even if activities have not been progressed
in strict logical sequence)

The number of users permitted

The amount and flexibility of data which can be attached to each
activity

The selection of a package will depend, of course, on the needs and
priorities of the user.

Cost-Estimating Software

It is interesting to note that far fewer cost-estimating packages are
available than scheduling packages. The reason for this is subject to
speculation, but it could be due to the fact that the approach taken
to cost estimating is less standardized, and it is therefore more
difficult to create a standard package which would satisfy many users.

There are a number of software packages available which are designed for specific types of estimating, such as architectural work. If
an available program suits the particular application, all well and good.
However, at the time of this writing, there will, most often, not be
programs that exactly fit the user's application. Fortunately, there
are a number of software packages available which are suitable for
general estimating applications. If we consider that cost estimating
usually consists of a lot of multiplication of quantities by unit costs,
and addition of the results, it is evident that it is a simple matter to
computerize it. The methods and data required for estimating can be
developed with available software as well. Some of the types of
programs that can be used are as follows.

"Spreadsheet" Software

Best illustrated by "Visicalc," "Multi-Plan," and "Lotus 1-2-3," spreadsheets enable the user to define a matrix of data as well as the calculations
required to define additional data. One can, for example, take a column
of quantities and multiply it by another column of unit costs to create a
detailed estimate. The author has found it useful to set up matrices
for detailed calculations which are then summarized and subjected to
further manipulation in a higher level matrix. Like that of most users
of spreadsheet software, the author's experience has been that the
more you work with it, the more things you find you can do with it.
Spreadsheet programs are generally used with microcomputers.

Database Software

Databases are one of the easiest and most useful computer tools available to the project engineer. Databases are discussed more fully after our discussion of economic analysis software. Cost-estimating applications are designed to collect, analyze, normalize (i.e., adjust to a consistent basis or "norm"), and present the basic data needed for estimating. Since the best predictor of future costs is our past experience, and since small projects continuously generate a lot of useful, actual cost data, a database is a handy way to organize it all.

Databases and spreadsheets can be integrated with the network in order to automate the estimating process described in Chapter 5, in which the labor and construction equipment resources are estimated from a standard file of rates. This makes it possible to examine the cost impact of different planning scenarios, as well as to make cost forecasts reflecting current progress.

A cost-estimating database should have an independent variable (e.g., design parameters such as cubic yards of concrete), and a corresponding value for the dependent variable (cost). It should be organized according to cost code, such that data can be easily selected, sorted, and summarized

Economic-Analysis Software

There are a number of software packages available which perform economic analysis in order to evaluate project profitability. These programs can be useful in establishing the return on investment, or in monitoring the projected return as variations to the original basis occur. It is worth noting that updates in the profitability analysis are seldom done after the project is approved, although they should be, as it is a relatively simple task for the project engineer to use an economic analysis to make decisions based not on minimum cost or schedule, but on maximum profitability.

Database Software

There are a number of database packages available, some of which are intended for use in conjunction with a network package. The database, as can be seen from the cost-engineering applications described above, enables us to collect, organize, analyze, sort, select, compare, and present any kind of data in any way we want. Some of the project-management applications of databases include:

Data on the durations of typical design and construction activities
Material control by storing and accessing data on purchase orders, purchase requisitions, delivery dates, quantities, etc.
Document control (i.e., keeping track of drawings, specifications, and critical communications)

Maintenance data (i.e., required inspection and maintenance
   intervals and procedures, and actual maintenance records for
   each item)
Weight data (for those projects in which weight is critical)
Personnel data
Wage rate data

Some database programs are like "blank slates" and can be designed
in any way that the user requires, while others are designed for a
specific application.

A very useful feature of some systems is the integration of a database with the network-analysis program. This feature enables us to
identify a resource requirement for an activity, and link that resource
to a detailed set of relevant data in the database. For example, we
might identify heat exchanger E103 as a resource requirement for the
activity "erect heat exchanger." By setting up a link to our database we can have the program select some or all of the data pertaining
to E103 and advise us of such facts as the vendor, the purchase order
number, the price, the promised delivery date, the current status, the
current delivery date, etc. We can even use this capability to update
our schedule automatically to reflect current deliveries, or to update
our cost forecast automatically to reflect current progress.

Management-Reporting Software

Management-reporting software consists of word processing, special
report-generating packages, and management graphics. This software
may be available within the existing computer system of the company,
or on word-processing equipment already installed. Features to look
for in these packages include the ease of use, compatibility with existing systems (such as word processing), and interface characteristics
with the main project-management programs. It is highly desirable
that the report-writing and management graphics packages either be
part of, or fully compatible with, the planning, scheduling, cost estimating, and project-control package(s) to avoid duplication of input
and assure timely reporting.

Risk-Analysis Software

For projects involving significant financial risks, a risk-analysis program may be helpful in addressing the probability of various costs and
schedules being achieved. For example, bidding strategy may nec-
cesitate an evaluation of bid price vs. the probability of getting the
job. Or a turnaround project which incurrs major costs for every day
the unit is shut down may benefit from a probabilistic analysis of
schedule to determine the probability that the unit will be running on
a given date, and to be able to plan accordingly.

Probabilistic analysis is appropriate for many more project-management situations than is generally acknowledged, probably because many engineers and managers are somewhat suspicious of statistics as a means of decision making. However, there are a number of software packages available which do a credible job of analyzing the probability of cost overrun or schedule delay (see Chapter 10). In general, these programs evaluate costs separately from schedules, and use a Monte Carlo simulation to generate the results. In Monte Carlo simulation, a random-number generator is used to select a value for each activity duration or cost variable, until a sample cost or schedule outcome is calculated. By repeating this process hundreds or even thousands of times, a probability function describing all possible outcomes can be developed. Some probabilistic packages which are well accepted are the "Range Estimating Program" (or REP) offered through McAuto, which is for probabilistic cost analysis, and "QGERT" which is a probabilistic schedule-analysis package. The major drawback to this type of analysis is that it requires that the variables be independent, which they seldom are.

If risk and contingency analysis is important, an appropriate software package can make the inevitable judgements more rational, credible, and acceptable. And, an appropriate package can point out the magnitude of risks which were previously thought to be inconsequential.

CURRENT TRENDS IN PROJECT-MANAGEMENT COMPUTER SYSTEMS

There are two basic trends observable today, relevant to those in the project-management field:

1. New hardware and software products and technology which can be used effectively in the project-management field are being introduced and being implemented constantly.
2. The recognition of project management as a business and engineering endeavor that requires computer assistance is also increasing.

As a result it is now considerably easier to select, justify, and implement a project-management system than it was even one year ago. In addition, these trends are continuing and even strenthening today. Some of the new products on the market or approaching the market include:

Terminals without keyboard: These systems use a small, handheld device known as a "mouse" to move a cursor on the screen and thereby input commands simply by pointing to a display of command options. Apple's Lisa and MacIntosh systems are the best known examples of this technology, in which the keyboard is required only for the entry

of data. Other designs, such as the Hewlett Packard HP-150, allow
the user to issue commands by touching the screen. "Voice recognition," another advance, is already available from Texas Instruments.

Multiple displays: These are systems which are capable of display-
ing the outputs of several programs in several "windows" at once. For
example, a project engineer could update his network and database,
while preparing a tracking curve, and see the results of these operations
simultaneously on the screen. Or, when preparing his progress report, he could use the word-processing, graphics, and spreadsheet
programs in an integrated mode to perform a calculation on the
spreadsheet, to plot the results, and to "paste" it into a report.

Integrated software: The latest crop of software packages for mini-
and microcomputers features a much higher level of integration of the
most commonly used computer functions. For example, programs such
as Lotus 1-2-3, symphony, and framework offer an integrated approach
to spreadsheets, graphics, and database operations. Other software allows the user to integrate programs of his choice, such as work processing, statistics, graphics, and financial modelling.

Integrated hardware: The current problems with matching hardware
and software components in such a way that they work well together
are gradually being solved. The lack of an industry standard and
the success of the IBM microcomputer, has made the IBM micro a de-
facto industry standard. Vendors in general are realizing that their
users expect to be able to do more than one task with their systems.

Local Area Networks (LAN): These local networks allow the various
computers, terminals, workstations, and central processing equipment
to communicate directly, thereby giving the user the power to utilize
the entire office system from his workstation. An example of LAN is
electronic mail, in which a memo or report prepared at one workstation
can be transmitted electronically to other workstations as well as be
printed for distribution. LAN are also used to link large and small
computers to make the added power of the larger machine available to
the micro-user. And specialized hardware, such as a printer, hard
disk, or plotter, can be utilized from several different workstations.

Machine speed and capacity: With the advent of 16-bit and 32-bit
microprocessors, low-cost hard discs, and other methods for increased
memory, the power and speed of minis and micros is rapidly increasing.
This, in turn, makes the individual user far more effective.

Organizational changes: As systems become easier to learn and use,
as well as more effective, their use is extending in all directions in
the organization. Managers who, a year ago, would have never thought
of using a computer themselves now rely on their own desktop model.
The same is true for many people at the working level. One notable
result of this is that the organizational structure for control of the
computing function is changing from the data-processing department
(designed to fit the mainframe computer) to a more integrated function,

embracing computing, word processing, communications, and data security.

MICROCOMPUTER APPLICATIONS TO SMALL PROJECT MANAGEMENT

Many project engineers were, for many years, in a frustrating situation in which they were aware of the need for computer assistance in their activities managing many small projects, but were unable to justify the high cost of effective systems. That all changed with the introduction of microcomputers with project-management software. Now, computer-assisted project management is within the reach of virtually every project engineer.

It should be noted that everything in this chapter pertains to micro-computers as well as to other types. However, since microcomputers are so relevant to the small-project environment, we discuss them specifically in these paragraphs.

The Microcomputer: A Practical Tool For Small Project Management

Several factors have recently combined with the result that micro-computers are finally a practical tool for the project engineer. These factors are:

The availability of a wide variety of microcomputer hardware with sufficient computing capacity and speed to be able to handle the data for one or more small projects.

The availability of micro-based project-management software which is easy to learn and effective to use

The low price of microcomputers

The growing ability to interface micros with other computers

The high level of awareness and acceptance of microcomputers as a practical business and management tool

The recent reductions the amount of time necessary to learn to use micros effectively

The growing acceptance of micros as "personal" computers in an organization, dedicated to an individual and/or task, thus giving the project engineer much greater freedom to explore possibilities and develop his own computer solutions

The growing variety of applications in the small-project environment

With all these trends, it is logical that microcomputers be considered for small projects. However, no computer can manage a project by it-self. It is therefore important to discuss the various aspects of micros, and point out their problems and pitfalls.

The Microcomputer Defined

Since it is currently very difficult to distinguish clearly between computer sizes and types, we will, for this discussion, define the microcomputer as a standalone system with the following characteristics:

48K to 512K of RAM (random-access memory)
Floppy disc storage with optional hard disc storage
A computer, visual display, keyboard input (optional mouse), disc storage, and a printer
A cost from $3000 to $10,000 (1984 prices excluding software)

Representative products include the Apple IIc, Lisa, and MacIntosh; the IBM PC,PC/XT,PC/AT; Texas Instruments PC; and Compaq. In this definition, we distinguish the microcomputer from the home computer and the smaller minicomputers. We use the terms microcomputer and personal computer interchangeably in this discussion

Microcomputer Configurations

Microcomputers can be configured as follows:

A system with input, processing, and output at one location.
A system with a local network to other micros and/or to the mainframe computer.
A link to more powerful storage device, such as hard disc.
A link to more sophisticated output devices such as four-pen plotters.
A data link for input and output to remote locations.

In other words, we can use the micro as a computer all by itself, or in conjunction with other facilities.

It is also interesting to note that many items of office equipment are now actually microcomputers. In the past, word processing equipment, telecommunications equipment, and computers were separate and distinct. Now, they are all forms of microcomputers. For example, some word processors are also microcomputers for which extensive software is available. On the other hand, software exists which will allow a standard micro to function as a telex machine, although modern telex machines are also a form of micro. The point is that micro capability can be found in a number of places. If certain functions need to be integrated, one can look, not only at the well-known micros, but at specialist equipment as well.

When to Consider a Microcomputer

For all the reasons just discussed, microcomputers are a viable solution in many project-management situations. However, there are some

specific situations where microcomputers are particularly appropriate
for project management:

When the mainframe computer is not available or cannot provide
timely project-management software in a distributed processing
format.

When the company is small with limited or no in-house computing
capability.

When timesharing usage is or is expected to be sufficiently costly
to justify purchase of a micro.

When computer services are required throughout the company but
a mainframe is considered too expensive.

By selecting the right configuration, micros can accomodate the project engineer handling up to about 20 projects at one time.

Pitfalls

As in any endeavor involving high technology, the user of microcomputers is apt to find that his actual experience turns out to be
quite different from what he expected. The author has served as a
willing "guinea pig" in this regard, by trying different products and
attempting different project-management operations on them, and the
resulting pleasure and pain gives rise to the following suggestions.

Marketing Microcomputers as a Consumer Product

One problem with microcomputers is that they are sold in a way that
resembles the sale of consumer products. This practice is commendable
in some ways but causes problems in the business and technical environment. One problem that results is a tendency of sales personnel
to be unfamiliar with the details of the hardware and software, as well
as unfamiliar with the business environment in which it will be used.
It is therefore important to select a dealer who is oriented towards
business applications, and to select a salesperson with specific experience with the type of hardware and software being sought. And it is
worthwhile to buy from a dealer who will be available for later advice
on additional hardware and software, and who is likely to provide
realistic, practical advice.

Failure to Assure Service After the Sale

Micros, like any technical device, may need servicing or repair. Some
dealers offer such services, but many do not. It is important to in-
quire as to service arrangements, particularly for business systems
which cannot be out of service for long. For this reason, on-site
service is generally preferred. It should also be noted that service

may be required from the software vendor as well, and the vendor's
reputation for supporting the product should be an important factor
in the decision to purchase.

Failure to Obtain Assistance When Getting Started

Unpacking and setting up a new micro is a bit like getting your first
stereo, as you are confronted with a bewildering array of manuals,
guidebooks, references, and instructions. Some dealers will set up the
system for you (e.g., install special memory cards, hook up the
printer, etc.). They may also give you free instructions in the basic
operations. This assistance is very valuable, and you help avoid the
"if I had only known!" type of complaint, as well as minimize the
amount of time required to get started.

Lack of Industry Standards for Compatibility

Perhaps the greatest problem facing the burgeoning microcomputer
industry today is the lack of an industry-wide standard for compatibility. This means that, although lots of hardware and lots of software
are available, much hardware will not work with much of the other
hardware, much software will not work with much of the other software, and a lot of the software won't run on a lot of the hardware.
The most practical way to avoid problems of compatibility is to "try
before you buy," even if it means bringing you own computer into the
store.

And, of course, it is essential to study the user manual thoroughly,
before purchasing hardware or software.

Fear of Obsolescence

Many people view the rapidly changing microcomputer scene and conclude that holding off buying a computer is the best course of action.
This will, they figure, result in a purchase later on of more computer
for less money, and avoid the chance that a computer purchased today
will become obsolete in a year or two. Indeed it is true that the cost of
computing will probably continue to drop, while the sophistication and
performance of the product will continue to increase. The system
purchased today may indeed, in some ways, be obsolete in a year or
two. So what are the reasons for getting started now?

Perhaps the most important reason is that, if we are convinced that
improved project management will enhance company profitability, we
can hardly afford to wait. We may save a few thousand dollars by
waiting before buying but may lose the opportunity to save tens of
thousands during that time. Our competitive edge may also suffer as
competitors begin to display the improved efficiencies we are deferring.
And, since micros are clearly here to stay, the sooner we learn how to
use them the better.

We should also look more closely at what we mean by "obsolescence." Even if the equipment we purchase today will be outperformed by new equipment next year, it doesn't mean that it will no longer be useful. As long as a computer is doing a useful job for us, it is not obsolete, even if it is not as fast or powerful as the latest models. Additionally, there are often new ways to upgrade existing systems to improve performance.

Applying Special Standards for
Computing Decisions

Because of the glamour and excitement surrounding microcomputers today, there is a tendency to bypass the usual company decision process on the grounds that computers are a special situation, and we just have to "get on the bandwagon" and buy one. In fact, there is no reason why a computer system under consideration, even if it is a micro, should not be subjected to the same evaluation standards that would be applied to any other capital purchase. And there is no reason why the systems design and selection procedures described in this chapter should not be adhered to as well. These procedures force a discipline on the decision process and help assure that the resulting system truly contributes to profitability.

COMPUTER-AIDED DESIGN APPLICATIONS FOR PROJECT MANAGEMENT

During the past several years, the trends toward greater computing power at much less cost have been very evident in the area of Comp-uter-Aided Design (CAD). As a result, both owner and contractor companies in many fields are finding that CAD offers great improvements in the efficiency with which design tasks are conducted. Therefore, project engineers and managers are becoming increasingly aware of and involved with CAD as a design and drafting tool. However, CAD systems also offer the potential for dramatic improvements in the project-management function as well. CAD systems are particularly relevent to small projects in a large operating plant which has an engineering function of sufficient size to warrant CAD systems.

Basic CAD Components and Capabilities

A CAD system is designed to perform many of the functions associated with the design process, including:

Drafting, including layouts, revisions, and final drawing
Storage, recall, and display of relevant design information
Material takeoffs for procurement
Modelling in three dimensions to remove interferences

Performance of design calculations
Application of design standards

To accomplish these tasks, a CAD system typically consists of the
following components:

A workstation consisting of one or more screen displays, and elec-
tronic table, a keyboard, and a device for locating points on the
screen or on the table (e.g., "mouse"-type cursors, light pens,
"joysticks," thumbwheels, etc.)
A central processing unit providing the hardware and software to
perform the necessary operations
Storage facilities such as tape or disc
Auxiliary input devices (as required)
Telecommunications facilities (as required)
Printers and plotters to prepare the drawings

In most cases, a single processing unit can support a number of
workstations, which also share the output devices. Some of the more
popular CAD systems are Computervision, Intergraph, Autotrol, and
CALMA.
As a result of the efficiencies introduced by CAD, both owner and
contractor companies have experienced improvements in design ef-
ficienciey by a factor of four or more. CAD systems are of interest,
not only because their use is becoming widespread, but also because
they have some features which are of great interest to the project engineering and management functions. Foremost among these features
is the database capability which enables us to attach management data
to design graphics. Current technology also embraces Computer-Aided
Manufacturing (CAM), in which the design of a part to be manufactured
can be translated directly to a tape for a numerically-controlled
production machine.

CAD Applications to Cost Control

One of the most common and long-standing challenges in cost engineering involves the exertion of cost control during the design phase.
Typically, this effort takes the form of procedures to control design
man-hours and costs, as well as to estimate and track design changes.
However, the most important aspect is usually not addressed, and that
is control of the cost implications of the many design decisions that
are made. This control is vital, because the design of the project is
the aspect over which we have the most control, and it is also the
aspect which has the most potential impact on cost.
Unfortunately, most companies are not very successful at integrating
design and cost engineering. This is partly because of the misplaced
belief that engineering work should be done in the minimum time and

cost. This, of course, discourages any discretionary work such as
efforts to reduce cost through design changes. Another problem is
that most design engineers have neither the time, data, nor inclination
to get involved in cost studies, just as most cost engineers are not
capable of design work.

The CAD system enables us to construct a database for estimating
the cost of each of the components included in the design. Thus the
cost of each alternative design configuration can be easily calculated
and provided to the design engineer. Changes can also be identified
and estimated more efficiently. Life-cycle costs can also be integrated
into the design process in this way.

CAD Applications to Weight Control

Many projects, such as offshore platforms, require monitoring and
control of weights. This can be done in a manner similar to that used
for cost control, by constructing a database of weight information attached to the graphical representation of each design component. The
resulting data and analysis of weights provides vital information for
platform structural analysis and planning of construction operations.

CAD Applications to Document Control

All projects have the problem of storage and organization of documents.
This is required not only for efficiency, but also for certification and
operation of the facility. Since many documents relate specifically to a
design component, the CAD system can be used to log in all documentation relating to a specific equipment or bulk-material item. Typical
documents stored in this way include:

Quality control documentation, such as testing and inspection results
Operating information
Vendor information
Maintenance information

CAD Applications to Maintenance

A typical problem of engineering and construction projects, large or
small, is the problem of "handling over" the completed facility to those
who will operate it. The operation personnel are interested in data
which was (presumably) accumulated during the engineering phase,
but which was of little interest to those who were designing and building the facility. Consequently, the operations staff often find it difficult to obtain and organize the information they need. One example
of this problem is maintenance.

In order to assure that an effective program for inspection and
maintenance is set up, we can use the CAD database to accumulate information relating to inspection and maintenance of each item shown on
a drawing. The database could show, for example:

The frequency of inspection
The intervals between planned maintenance
The procedures to be followed for planned and unplanned
maintenance

Once the facilities are in operation, and the maintenance program is
in effect, additional information could be added to the database to
facilitate planning and monitoring of the inspection and maintenance
function. Additional information that could be added includes, for
example:

The date of last inspection
The result of last inspection
The date for the next inspection
The date of the last planned maintenance
The maintenance performed
The date of the next scheduled maintenance
The maintenance to be performed

CAD Applications to Construction Planning

Most CAD systems have the capability to work in three dimensions.
This capability makes it possible to model the construction work to be
done and see potential interferences, misfits, or impractical construction requirements. For example, one of the uses of a plastic model is
to be able to actually see, in three dimensions, what is going to be
built. The three-dimensional model on a CAD system not only lets us
see the project in three dimensions, but also enables us to "move things
around" with ease. This feature can be used to test the feasibility of
different arrangements, and to actually simulate construction operations
such as major lifts. For small projects in dense operating plants, the
three-dimensional modelling feature can be a big help in avoiding expensive situations in which the design looks good on paper but proves
to be impossible to construct in the field.
    Finally, by simple speeding up the design process and the production
of approved drawings for construction, the CAD system can do a lot
to make construction go more smoothly.

CHAPTER SUMMARY

In this chapter we reviewed the availability of computer tools for project-management applications. We saw that there are many computer
systems that are appropriate for small projects. We also discussed the
necessary steps for system design, specification, selection, and
implementation.

Index